THE DYNAMICS OF NAZISM

Leadership, Ideology, and the Holocaust

STUDIES IN SOCIAL DISCONTINUITY

Under the Consulting Editorship of:

CHARLES TILLY
University of Michigan

EDWARD SHORTER
University of Toronto

In preparation

Paul Oquist. Violence, Conflict, and Politics in Columbia

Richard C. Trexler. Renaissance Florence: The Public Life of a Complex Society

Samuel Kline Cohn, Jr. The Laboring Classes in Renaissance Florence

Published

Fred Weinstein. The Dynamics of Nazism: Leadership, Ideology, and the Holocaust

John R. Hanson II. Trade in Transition: Exports from the Third World, 1840–1900

Evelyne Huber Stephens. The Politics of Workers' Participation: The Peruvian Approach in Comparative Perspective

Albert Bergesen (Ed.). Studies of the Modern World-System

Lucile H. Brockway. Science and Colonial Expansion: The Role of the British Royal Botanic Gardens

James Lang. Portuguese Brazil: The King's Plantation

Elizabeth Hafkin Pleck. Black Migration and Poverty: Boston 1865-1900

Harvey J. Graff. The Literacy Myth: Literacy and Social Structure in the Nineteenth-Century City

Michael Haines. Fertility and Occupation: Population Patterns in Industrialization

Keith Wrightson and David Levine. Poverty and Piety in an English Village: Terling, 1525-1700

Henry A. Gemery and Jan S. Hogendorn (Eds.). The Uncommon Market: Essays in the Economic History of the Atlantic Slave Trade

Tamara K. Hareven (Ed.). Transitions: The Family and the Life Course in Historical Perspective

Randolph Trumbach. The Rise of the Egalitarian Family: Aristocratic Kinship and Domestic Relations in Eighteenth-Century England

Arthur L. Stinchcombe. Theoretical Methods in Social History

Juan G. Espinosa and Andrew S. Zimbalist. Economic Democracy: Workers' Participation in Chilean Industry 1970-1973

Richard Maxwell Brown and Don E. Fehrenbacher (Eds.). Tradition, Conflict, and Modernization: Perspectives on the American Revolution

Harry W. Pearson. The Livelihood of Man by Karl Polanyi

The list of titles in this series continues on the last page of this volume

943.086/WEI

THE DYNAMICS OF NAZISM

Leadership, Ideology, and the Holocaust

FRED WEINSTEIN

*Department of History
State University of New York
Stony Brook, New York*

Academic Press

A Subsidiary of Harcourt Brace Jovanovich, Publishers
New York   London   Toronto   Sydney   San Francisco

COPYRIGHT © 1980, BY ACADEMIC PRESS, INC.
ALL RIGHTS RESERVED.
NO PART OF THIS PUBLICATION MAY BE REPRODUCED OR
TRANSMITTED IN ANY FORM OR BY ANY MEANS, ELECTRONIC
OR MECHANICAL, INCLUDING PHOTOCOPY, RECORDING, OR ANY
INFORMATION STORAGE AND RETRIEVAL SYSTEM, WITHOUT
PERMISSION IN WRITING FROM THE PUBLISHER.

ACADEMIC PRESS, INC.
111 Fifth Avenue, New York, New York 10003

United Kingdom Edition published by
ACADEMIC PRESS, INC. (LONDON) LTD.
24/28 Oval Road, London NW1 7DX

Library of Congress Cataloging in Publication Data

Weinstein, Fred
    The dynamics of Nazism: Leadership, ideology, and the Holocaust.

    (Studies in social discontinuity)
    Includes bibliographical references and index.
    1. Hitler, Adolf, 1889–1945. 2. National
socialism––History. 3. Holocaust, Jewish (1939–
1945)–Germany. 4. Germany––Politics and government
––1918–1933. 5. Germany––Politics and government
––1933–1945. I. Title. II. Series.
DD247.H5W38   943.086   80–514
ISBN 0–12–742480–6

PRINTED IN THE UNITED STATES OF AMERICA

80 81 82 83   9 8 7 6 5 4 3 2 1

FOR THE CHILDREN

Blaine
Malcolm
Robert
Alison
Samuel

Contents

Preface ix
Introduction xiii

1
The Sense of Continuity and the Common Sense World   1

Nazism and the Conservative Intelligentsia   1
Some Examples of Conservative Support and Withdrawal   6
The Conservative Intelligentsia and the Common Sense World   15
The Routine Interpretation of Reality in Common Sense Terms   21
Some Critical Comments on Social Scientific Views of Nazism   29

2
The Emotional Impact of Events:
When the Common Sense World Fails   47

The Importance of the Concept of Affect
in Theory and History   53
The Psychosocial Bases of Hitler's Appeal   57
The Problem of Continuity   70

3
Hitler's Charismatic Power and Racism 79

Some Comments on Traditional Psychoanalytic Orientations
to Social Action 104

4
Nazism as an Ideological Movement: Conceiving the Final Solution 117

Orientations to Past and Future 122
Orientations to Struggle and Violence 126
Orientations to Experience and Affect 128
Orientations to Hierarchy 132
Conceiving the Final Solution 136

Index 163

Preface

The burgeoning literature on Nazism has not much enhanced our understanding of the subject. Responsible scholars have produced a great deal of empirically rigorous work, and objective data of all kinds have been made available. Still, one cannot infer from such data what Nazism represented to the German people; so interpretations, even those of the best of writers, are unconvincing, and no one is satisfied that any but the most elementary sort of objective problem has been solved.

The dissatisfaction with the literature is understandable, but it would be a mistake to be affected by it; the interpretive impasse can only be broken by further work. Indeed, as long as the inner significance of Nazism remains a riddle, as long as the Holocaust particularly remains unassimilated in imagination, incomprehensible in either common sense or theoretical terms, another attempt to shed some light on the experience, to see whether and to what extent we might possibly develop an approach and a language adequate to an interpretation of it, cannot be amiss.

Obviously, such a project is more easily contemplated than accomplished. No succession of traditional narratives can serve its fulfillment; nor can fiction serve, even though fiction harbors its own truth. At the same time, the empirical reality raises serious problems for alternative theories. Sociologists and social historians have addressed themselves many times to the Nazi movement, seeking for the unities that presumably characterize such a movement, trying to identify a common cause for a common effect, employing familiar categories of analysis—such as class—with the aim of establishing statistically significant relationships. [But the interesting thing about the Nazi movement is the heterogeneous support it mustered, sufficient to compromise the validity of any inference that might be drawn from such established relationships, primarily because the approach itself, the mode of analysis, has nothing to offer with respect to the subjective strivings and perceptions that led people to support the movement or to act on its mandates.]

Moreover, psychoanalysts and psychohistorians, whose business it is to explain subjective strivings and perceptions, have not successfully done so, either because they have tried to solve a complex problem by resorting to one or another "id mythology" (more readily invented than believed) or because they have followed the conventional sociological approach, converting psychoanalytic insight into objective categories—relating class to familial socialization and hence to character, or age to radical enthusiasm—and then searching for the numbers by way of confirmation. Psychoanalytically derived categories, however, have been even less helpful in accounting for the heterogeneity of movement support than the sociological categories they were supposed to augment, or even replace. The explanations of perceived relationships in these terms have either been so contradictory or have required such inferential leaps that no empirically oriented historian or sociologist can accept them.

These brief comments help to explain why our understanding of Nazism has been so little advanced. Approaches rooted in conventional sociological thinking cannot encompass the experience because they are inappropriate to an explanation of the heart of that experience—the extravagant and stunningly arrogant brutality of the Nazis. At the same time, psychological and psychoanalytic approaches, specifically meant to provide such explanation, have nevertheless failed to meet the standards of empirical adequacy, especially the standard of verifiability, which serves as an obligated check against unwarranted conclusions.

This book focuses on problems of theory as well as of history and attempts to explore the ways in which the basic concern of [psychoanalysis (subjective strivings and perceptions) might be integrated with the basic concern of sociology (organized collective behavior)] The extent to which we can ever actually understand Nazism as a historical phenomenon depends upon the extent to which we can achieve such an integration, that is, encompass the subjectivity of the people involved in an empirically adequate way, on some level other than the biographical.

I could not have undertaken or completed this book without the constant help of my friend and colleague, Gerald Platt. No simple acknowledgment can express the extent to which I have been instructed by Platt's thinking about sociology and psychoanalysis. His ideas are so heavily represented in this book that in a significant sense it is also his.

I have also relied consistently on contemporary psychoanalytic thought, particularly on the work of Joseph Sandler, whose ideas clarified for me the way in which ideology works to stabilize social activity over time and how people are likely to respond when ideological commitments are rendered problematic by disruptive social conditions.

A number of my friends and colleagues read the manuscript and made valuable suggestions, and saved me from more errors than I am likely to remember, including Mel Albin, Werner Angress, Ruth Cowan, Cornelia Levine, Robert Levine, Bruce Mazlish, and William R. Taylor. Among the many other people who helped me in important ways, I must mention Konrad Bieber, David Burner, Walter Laqueur, Gene Lebovics, Beth Lewis, Timothy Mason, the late Benjamin Nelson, Neil Smelser, Nancy Tomes, and Robert Wistrich. I am indebted as well to Gita Johnson, Janet Langmaid, and Christa Wichmann of the Wiener Library, London; to their coworkers, Richard Diamond, Margaret Rosenthal, and Oren Stone; and especially to Noel Mathews, also of London. I could not have accomplished as much as I did during my sabbatical year without the help of these generous people.

Finally, I must also thank the American Council of Learned

Societies, the Lewis Rabinowitz Foundation, and the Memorial Foundation for Jewish Culture for their generous financial assistance in the completion of this work.

I have quoted with permission from Erich Matthias and Rudolf Morsey, eds., *Das Ende der Parteien* (Düsseldorf, 1960), pp. 239–240; Rolf Seeliger, ed., *Braune Universität: Deutsche Hochschullehrer gestern und heute,* 6 vols. (Munich, 1966), 3, pp. 49–51.

Introduction

Adolf Hitler decided that a purified racial community and the passionate commitment to discipline, authority, and hierarchy it required could never be achieved in the context of traditional conservative politics. Demands for radical change from the right could not command popular support given the enduring caste and class exclusiveness of aristocratic landowners, army officers, bankers, and industrialists. Hitler urged rather the adoption of the fundamental strategy of the despised Marxist socialism—the organization of a mass movement.[1] Only a mass movement bound by racial and national loyalties could overcome the divisive effects of class conflict, the seemingly endless proliferation of interests, moralities, and parties.

The novelty of Hitler's solution to the dilemma of right-wing politics in Germany, the organization of a mass movement, was underscored by the novelty of his own status as a political figure. Hitler repeatedly referred to himself as an anonymous individual, an unknown front-line soldier, just one of eight million such soldiers, who, bolstered by the force of his own convictions, had driven himself and had risen from obscurity to become a great leader.[2] Hitler held himself up before others as proof that defiant, heroic individuals could compel a radical departure from the taken-for-granted course of events by sheer force of will.

Hitler saw his mission and his instrument, the Nazi movement, as unfolding outside the mainstream of German history—and why not? Who would have believed that an Austrian, serving in the German army with courage and distinction but still never rising above the rank of corporal, might become the leader of a great mass movement and achieve power in *that* society, with its constant stress on class distinctions and social background. Hitler risked himself and won. But he was able to do so because he did not believe in the finality of social, particularly class, relationships, or in the irreversibility of social change. Indeed, Hitler was confident of his ability to forge a unified racial community because he saw his own life as a repudiation of the effects of class, which he had subordinated to his will.

The desire for a homogeneous, racially harmonious community based on a peculiarly German but nonetheless integrative version of socialism and on revolutionary deeds carried out in its name was commonplace among right-wing intellectuals and activists in postwar Germany. There was considerable debate over the meaning of the concepts of race, nation, and especially of socialism as it implicated the national or racial community.[3] But Hitler had arrived at his own solution: Confirming his sense of Nazism as a radical departure from a failed bourgeois past, Hitler offered a unique vision of the virtue and necessity of perpetual warfare, of ruthless movement through space to secure by conquest the future of the racial community, which he thought could be rendered impervious to the effects of time.

Hitler's vision, based on an unmatched grandiosity and ferocity, inspired the Holocaust—unarguable evidence for the uniqueness and peculiarity of the experience. Neither Hitler nor the Holocaust was prefigured in Germany's past: On the contrary, no one anticipated the extraordinary consequences of Nazi activities because no one imagined the possibility of racially animated, systematic, technologically proficient mass murder of defenseless people in violation of every commonly held cognitive and moral standard.

The familiar and opposed idea that Nazism was the logical,

inevitable, or necessary outcome of German history bears little relationship to these events. But that is characteristic of the principal ideas about Nazism, which are more likely to be ideologically consoling than historically accurate, more useful for restoring a sense of order to a disordered world for the sake of those who followed than for describing reality as it was perceived or experienced by the participants.

Consider, for example, the equally familiar description of the personality type that supposedly proved susceptible to Nazi appeals in the crisis years 1930–1933.

> [The] lower class individual is likely to have been exposed to punishment, lack of love, and a general atmosphere of tension and aggression since early childhood—all experiences which *tend* to produce deep-rooted hostilities expressed by ethnic prejudice, political authoritarianism, and chiliastic transvaluational religion. His educational attainment is less than that of men with higher socioeconomic status, and his association as a child with others of similar background not only fails to stimulate his intellectual interests but also creates an atmosphere which prevents his educational experience from increasing his general social sophistication and his understanding of different groups and ideas. . . . [A] low level of sophistication and a high degree of insecurity predispose an individual toward an extremist view of politics.[4]

This description of behavioral predispositions to authoritarian solutions characterized by anger, envy, and ignorance—the result of familial, intellectual, and cultural impoverishment—is consistent with the prevailing sociological view of Nazism as a lower-middle- and middle-class phenomenon. The description is an "ideal–typical" one, of course, and is meant to account for the most general features of the greater number of people and not in any way all the features of any single person or of all the people. The description is organized in these terms, however, in an attempt to bypass the main problem presented by the data: The German population was not psychologically homogeneous at all, in class or in any other terms.[5]

The description is based, moreover, on a series of unwarranted logical leaps that require us to infer shared subjective strivings from shared location in society, or shared motives from shared behavior.⁶ This ideal–typical description of German character is literally a fiction necessitated by the apparent inability to explain how a dynamically heterogeneous population was mobilized in a relatively stable movement over time. Finally, even if there was some basis in reality for this description in 1933, there is still no way to account in the same terms for the Holocaust, the genocidal activities of 1942–1944, without making other logical leaps.

The Nazi experience, as Hitler often said, was improbable. Understanding how Nazism succeeded, how the improbable came to pass, has little to do with such conventional conceptions of personality and society. These conceptions force exaggerated attention on objective factors and on continuity of experience, particularly in the analysis of the Nazis' seizure of power, despite the often-reported and well-documented sense of radical discontinuity, a subjective experience that obviously made Nazism a plausible alternative to large numbers of people.⁷ At the same time, of course, objective factors and the notion of continuity are virtually irrelevant to an explanation of the Holocaust, which is likely to be treated, by contrast with the seizure of power, as something mysterious, unfortunate, and incomprehensible. The beginning and end points of Nazism, rather close in time and involving basically the same people, are treated, in other words, as if they had emerged from two different worlds and as if there could be no empirically valid terms by which the two might be linked.

Nazism obviously merits our continued attention, but it is equally obvious that conventional approaches are inappropriate to an explanation of what we really want to know: the reasons, the intentions, the motives, the subjective strivings of people who ultimately proved responsible for such a unique and extraordinary degree of destructiveness. On principle, we have learned as much about Nazism as we can from numerate, objective studies of class, age, or occupation; from analyses of voting records; and from ideal–typical descriptions of German character, which are used primarily to hold the idea of character constant, that is, to ignore it, while objective problems continue to be studied. Clearly, we cannot gain access to the subjective world of the participants in these terms. We must therefore change the terms, pose the problem in a different way, and try to understand how it was that Hitler and the Nazi movement, inspired by and promoting a particular kind of ideological commitment, could have intervened in what was, from the standpoint of large numbers of people, an unexpected, unanticipated, potentially chaotic social situation and have unified and mobilized a heterogeneous population from the early 1930s through the war and the Holocaust to the final crushing defeat. This problem, involving the analysis of subjective strivings and subjective perceptions, is the most difficult kind of problem to deal with. Still, we are obligated to try, for to do otherwise at this point, to pursue conventional lines of argument, is to pursue a hopelessly ideological course. *

Notes

1. Hermann Rauschning, *Hitler Speaks* (London, 1939), p. 185.
2. This theme of the anonymous, nameless individual recurs often in *Mein Kampf*, trans. Ralph Manheim (London, 1969), pp. 171, 188, 203, 320; it turns up routinely in speeches (e.g., in to the First Congress of German Workers, May 10, 1933, in Norman H. Baynes, ed., *The Speeches of Adolf Hitler*, 2 vols. [London, 1942], 1:862). Hitler's achievement in these terms was impressive, and the theme was often reported by people who heard it. See Werner Jochmann, ed., *Nationalsozialismus und Revolution: Ursprung und*

---

* The concept of ideology is employed throughout this book in the sense intended by Marx and Engels: "Ideology is a process accomplished by the so-called thinker consciously, it is true, but with a false consciousness. The real motive forces impelling him remain unknown to him; otherwise it would simply not be an ideological process."[8] I am suggesting at this point that one of the social functions of historical work, perhaps the primary one, is simply ideological, that is, to link past, present, and future, to supply in consistent terms an indispensable sense of the world as continuous and comprehensible. Nazism presented particular difficulties in this regard, and when historians could not find appropriate terms, they borrowed terms from the social sciences, which, bourgeois or Marxist, have a similar function. It then becomes necessary to see the extent to which past work on Nazism is ideological and the extent to which analyses of Nazism can have a nonideological, "objective" purpose. There are other uses of the ideology concept, which is discussed in Chapter 4.

Geschichte der NSDAP in Hamburg 1922–1933. Dokumente (Frankfurt a/M., 1963), p. 405 (from the diaries of Luisa Solmitz, April 23, 1932); Erich Ebermayer, *Denn heute gehört uns Deutschland . . . Persönliches und politisches Tagebuch* (Hamburg, 1959), p. 200 (November 6, 1933). Frederick Böök, *An Eyewitness in Germany* (London, 1933), pp. 63, 66 (May 1933). See also Hans-Jochen Gamm, *Der braune Kult* (Hamburg, 1962), pp. 25, 29–30. This theme is part of the narcissistic tendency undergirding Hitler's thinking, to be discussed at length in Chapter 3. Its significance is discussed by Sheldon Bach, "On Narcissistic Fantasies," *International Review of Psychoanalysis*, vol. 4, no. 3 (1977):291.

3. Kurt Sontheimer, *Antidemokratisches Denken in der Weimarer Republik* (Munich, 1962), pp. 280–282.

4. Seymour Martin Lipset, *Political Man: The Social Basis of Politics* (New York, 1963), p. 114. Italics added.

5. The movement was heterogeneous in social composition too. I will discuss the psychological and social heterogeneity at length, in Chapters 1 and 2.

6. The large and ever growing number of objective, statistically rigorous, local, regional, and national studies of the relationship of class, age, occupation, region, and religion to the Nazi movement typically include or imply some common sense notion of psychological strain, of how people in particular economic or social situations are likely to feel or to respond. But there is no way to infer subjective motives from objective location in society or from common sense notions of psychological strain. Clearly, scholars reproduce this kind of study because it is feasible and not because on principle we are likely to learn anything more than we already know.

7. The cultural and intellectual interpretations of Nazism, based typically on the abiding presence of a volkish conception of a peculiarly worthy historical or racial German community, and the militant repudiation of modernity as a means of preserving this community, can easily be dismissed in these terms. Lacking any systematic conception of social and psychological processes, such interpretations have always rested on a confusion of content with process. True, there was a language characteristic of German conservative or reactionary thought, but what counts is not *what* was believed, for the contents associated with romantic, volkish, racial, and national thinking were not cumulative, constant, or homogeneous over time. What counts rather is the process, that is, *how* a fragmented, diverse set of beliefs was transformed into a relatively stable basis for action, how a particularly disruptive period affected people so that certain attitudes, ideas, and strivings became focused and legitimated, leading finally to the revaluation of cultural standards.

8. Howard Selsam, David Goldway, Harry Martel, eds., *Dynamics of Social Change* (New York, 1975), p. 72.

THE DYNAMICS OF NAZISM

Leadership, Ideology, and the Holocaust

1
The Sense of Continuity and the Common Sense World

Nazism and the Conservative Intelligentsia

In a speech of September 5, 1934, Hitler resolved a dispute between rival factions struggling to influence the future of artistic and cultural life in Germany. Hitler repudiated both factions, the volkish, which he thought was backward looking and sprung from a fuzzy world of romantic, "teutonic" conceptions, and the modernist, which he thought was contaminated by a Jewish spirit and never to be taken seriously, requiring no more to be dismissed than an ominous growl. Hitler favored representational clarity in art—ideal images of manliness, resolve, power, and beauty, which he thought should serve as a standard for people, encouraging them to believe in a grander future.[1]

The adherents of volkish art had made a mistake in thinking that they would be allowed to dominate the cultural life of the nation, though they might well have believed that Hitler shared their views or was sympathetic to them. But the modernists had made an even greater mistake, and a much more interesting one, considering their emphasis on energy, prophesy, and ecstasy in art, on movement, on a dizzying "chaos of the soul," an explosive, experimental dissolution of reality, all so remote from the philistine conceptions typically associated with Nazism.

The modernists wanted to represent in art what they conceived of as a genuine revolutionary impulse in Nazism. There were not many of them, to be sure: The cultural avant-garde of a radical Nazism was not likely to appeal to many people. Even so, they hoped to have an effect; they had their public defenders and spokesmen, and their sympathizers in government as well. Modernism was a tendency within the Nazi movement and an inexplicable one, it would seem, from any conventional point of view.

Expressionism was the most important of the modernist perspectives, and it was represented pre-eminently at the time by Gottfried Benn, a practicing physician, one of the outstanding German poets of this century, and a man possessed of extravagant racial and artistic visions. Benn enthusiastically greeted the Nazis' accession to power in 1933, believing that they represented not merely a new regime but a new vision of the birth of man "pouring floods of ancestral vitality through the eroded spaces of Europe."[2] Benn described expressionism as the last great resurgence of art in Europe, and the Third Reich as perhaps the last great conception of the white race, heralding "the emergence of a new biological type, a change of direction in history, the coming of a new national stock."[3] He expected Germany to be transformed ethically and esthetically by a novel reconciliation of delirium and discipline.

In a radio address of April 24, 1933, the poet spoke of the grand new state that would dissolve the sterile Marxist opposition between workers and employers, and he also spoke to the youth, complimenting them on having rejected the traps and fetishes of a defeated intelligentsia.[4] Benn was so convinced of the potentialities of Nazism in 1933 that he consistently defended the aspirations of the movement as he understood them, including the eugenic emphasis, especially against the exiles and dissidents.[5] He had no patience with a rationally ordered world; he had to choose, as everyone did, and he chose Nazism because it was elemental and impulsive.

Benn argued that no really great epoch in history had ever interpreted the human essence as anything but "irrational," which meant to him a capacity for creative activity. Bourgeois conceptions of history were incorrect and inappropriate: The great transformations in history were never the result of free discussions followed by elections; they were compelled by forceful people, the result of a combination of authoritative power and high culture. This was true for politics as well as for art. Besides, Benn said that he had to declare himself for this new state because he belonged to the people who gave it life. He had his own aspirations for this community and his own conception of what was occurring. But even if his aspirations were not realized and his conception proved wrong, that community would still be his.[6]

Benn initially saw this Nazi revolution (and life) as process, an eruptive unfolding of potentiality, unpredictable and open-ended, not fixed or rationalized but mastered, if at all, by artistic form.* Benn elevated feeling, action, and body over reason, contemplation, and mind as a means of breaking down artificial barriers created by the orderly, organized, technological world—a variation of the criticism of both Western liberalism and Marxism standard for the time. It was just not illuminating, as far as Benn was concerned, to be involved in commercial pursuits or to be submerged in a universally inclusive socialist movement: There was no valor in the one and no distinction in the other. Benn opposed specialization, abstract relativism, pragmatism, the ceaseless absorption of subjective strivings through the soulless, mechanical means of electoral majorities, and hoped that 1933 was the beginning of an entirely new era.

There was no reason, then, why parliamentary democracy (and psychological and intellectual individualism) should not be superseded on behalf of heroic strivings in the service of the race. Benn welcomed the power of the great leader whose appearance he held to be a spiritual principle and a recurrent phenomenon in history, especially in catastrophic times. He also welcomed the idea of dictatorship because it promised to reconcile economic and political conflicts, and people would no longer have to contend in these spheres.[8]

* Benn had earlier concluded with respect to the Bolsheviks that "he who would organize life will never create art, nor can he consider himself belonging to art."[7] Benn thought the Nazis represented something else.

Benn was preoccupied with form, with the ability of the artist to wrest form from potential chaos. He spoke of the absolute poem, without faith or hope, addressed to no one but assembled in a fascinating way, celebrating the morality of form—the profound technical mastery that art demands.[9] Content betrays the artistic impulse: Thought is always the child of want, words plead some special case. Benn wanted an art abstracted from meaning, without moral messages, an art that did not express ideals and one that could not in any way be absorbed in an exploitable psychological process.[10]

His idea of form as an absolute he at first imagined to be consistent with the Nazis' concern for breeding, selection, and training.[11] Benn expected Hitler, the heroic leader, to open the world to constantly emerging, ever-changing potentialities. He thought that the Nazis' constant stress on dynamic, restless activity and vital energy portended a radical breakthrough to new forms of life,[12] that Nazism represented a turning point for man and nature, that they would yet rescue something from a projected decline of the white race. That which lives, he wrote, is different from that which thinks, and insofar as that implied something elemental and primitive, it was also better.[13] Benn thought, as Hitler did too, though obviously for other reasons, that the modern commitment to rationality and intellect had to be transcended, that what would disappear was just an uncritically accepted historical accretion, and that man would thus be freed to realize other potentialities.

Gottfried Benn despised philistine conceptions of art and mechanistic conceptions of life, and yet he failed to see that the Nazi leadership wanted both, no matter how often they spoke about dynamic, creative, heroic life. Nothing could have been further from Benn's aspirations than Hitler's vision of endless ranks of marching blond youth who must appear as if they were hatched from the same egg. But Benn missed it; he had seen another thing altogether.

Benn's excited hopes for Nazism did not last long. Hitler had no tolerance for anything that Benn represented, and the Nazis attacked expressionism as debased, anarchic, selfish, snobbish, another form of cultural Bolshevism, and offensive to the healthy instincts of the people.[14] They attacked Benn personally, inferring contamination by Jewish blood from his art (forcing him to demonstrate, by way of response, that his name was of Wend, not Hebrew, origin).[15] Some 18 months after the Nazis had come to power, Benn concluded that the regime was a "horrible tragedy."[16]

It may seem remarkable in retrospect that a poet of Benn's ability and stature actually expected the Nazis to allow artists a free creative space, or that his racial visions and conception of life had anything at all to do with theirs. It may seem remarkable, too, that a number of other artists and their followers saw in modernist perspectives a revolutionary style appropriate to Nazism, or that a significant number of traditional, conservative artists—not all of them volkish—were willing to serve the movement though they had still other views of it.* The fact is, however, that Hitler had ample support upon his accession to power from the artistic as well as from the intellectual and professional communities, and that many well-educated, sophisticated people, representing a variety of tendencies, expectations, and beliefs, were quite prepared to serve.[17]

This support did not come just from idiosyncratic individuals, nor was it merely opportunist, though it was often just that.[18] Rather, the conservative and nationalist intelligentsia were disposed to construe the situation favorably; many among them saw in Nazism the elements of a genuine moral (or as they would have phrased it, spiritual) regeneration. Thus, the German public was bombarded from every side with the most exalted statements in praise of Hitler and the Nazi movement,[19] bolstering the sense of exhilaration as expressed in ideas of renewal, rebirth, rejuvenation, the superiority of the German people and culture, the elevation of instinct, will, and intuition in deliberate repudiation of Western orientations to rationality, thus enhancing a grandiose sense of moral worth that could only appear in retrospect as unbridled arrogance. These statements came from academics and artists, as well as from theologians, jurists, doctors, and others who, whatever their personal ambitions may have been, justified for any number of people a release from traditional constraints

---

* There were still other artists and intellectuals not interested in Nazism, radical or otherwise, who expected that the regime would at least allow them to work. Many intellectuals did of course continue to work, but uncommitted artists tended not to be so fortunate.

leading—among other things—to the book burning episode of May 10, 1933.

Hitler's assumption of power electrified a significant portion of this academic–intellectual–professional community at the moment, giving rise to enthusiastically expressed convictions that the turmoil of recent years was at last ended, that a rescue operation had been completed. The sense of relief and the expressions of gratitude were remarkable for their intensity.

The sense of relief was based on a conventional conception of the uses of political power, that is, on the expectation that Hitler and the Nazis would permit the expression of a variety of conservative and nationalist tendencies. Hitler, however, had no intention of using power in a conventional way, nor did he intend to permit the expression of competing tendencies. Every Nazi policy decision, therefore, required these people to decide in turn whether to continue to support the movement or not, and many of them, like Gottfried Benn, withdrew their support and at least privately repudiated the movement.

The ability of these people to identify with and participate in the Nazi movement (for however brief a time), the diversity of aspiration, striving, and insight they expressed, the differences in their character and flexibility of judgment, the desire of some of them to get out sooner or later, while others of similar background continued to serve, all this is of interest as it demonstrates that the image of a monolithic, totalitarian enterprise, pursuing a single-minded course to a known end, is no more applicable to the mental activity of individual Germans than it is to the bureaucratic administration of the Nazi state. It is particularly interesting that these people cannot be accommodated to any prevailing characterological conception of what Nazis were like.

Some Examples of Conservative Support and Withdrawal

Martin Heidegger's personal profession of faith in Hitler and the Nazi movement is indicative of just how affecting the Nazi seizure of power was. Heidegger was quite emphatic and laudatory, and he went to considerable lengths to exalt the new Germany, pledging unconditional loyalty to Hilter as the savior of Germany from its "suffering, disunity and forlornness," as the man who would lead Germany to unity and honor. Heidegger celebrated Hitler as the inspiring teacher of "a new spirit," and he told his students at one point that Hitler would liberate them: "The Führer, himself and alone, is Germany's reality and law, today and in the future." [20]

Commentators have often treated Heidegger's commitment to Nazism as an unfortunate, perhaps trivial episode in the life and work of one of the pre-eminent thinkers of this century. Hannah Arendt, for one, urged us to consider the timeless power of Heidegger's thought rather than this momentary lapse from which he shortly recovered.[21] This, however, is a mistake, and whereas it is not important to examine Heidegger's commitment in detail here, it is important to observe how easily it flowed from his personal inclinations, his philosophical thought, and the cultural preoccupations of his time.

Heidegger's vision of man thrown into the world alone and afraid should be taken as confession before it is taken as insight. That is, Heidegger's concern with being was remarkable in its focus on what people *are* rather than on what people *do*. Even Freud's great introspective and analytic capacities were focused primarily on impulse, transgression, and punishment (the things that people actually or wishfully do—the failures that make them neurotic), rather than on action or events that threaten disintegration of self (implicating what people are—the failures that make them psychotic).[22] People do not ordinarily reflect on how it is that they are, or can continue to be, until they are threatened by or otherwise achieve insight into madness, the disintegration of self. Heidegger's stress on anxiety stemming from the need to move constantly into a future in which the most significant and dramatic feature is death and his focus on mastery of self, on the willingness and the ability to create meaning, all originated, it would seem, from this kind of insight, the result of an introspective quest.[23] *

---

* Heidegger's insight reflects in the first place a fear of madness, not death, as people who are alone and afraid must constantly fear the disinte-

Heidegger's insight led him to Nazism, as the movement posed itself as a defense against any further violations of German honor and pride, and as a source of support and guidance for people who had lost their sense of integration with the world, who were threatened in their security of being, their sense of self. Nazism was important to Heidegger because it promised to develop in a modern, technological context the values of discipline, sacrifice, service, community, and authority. Nazism promised rootedness, which was missing from both the superficial, self-seeking, technical–rational orientations of liberalism, and from the internationalist, leveling, technical–rational orientations of Marxism."[4]

This insight of Heidegger's into ontological insecurities is important—no matter how grievously he misjudged Nazism—because it underscores the central German dilemma at the time and says quite a lot about how Germans felt and what they wanted. And it is interestingly related to Freud's late revision of his theory of anxiety, which also suggested the centrality of ontological insecurities stemming from life in society. In either case, it is not only a question of what people desire but also of what they fear, not only with respect to drives, which Freud was particularly interested in, but with respect to the environment. For people can hold themselves together only as long as they can perceive society holding together; they will do whatever they can when society appears threatened, with whatever degree of cognitive control is left to them, to maintain or restore a sense

---

gration of self. My conclusion here is speculative, and any more definite a statement must await the publication of private documents. However, it is a melancholy fact that Heidegger never commented on the Holocaust, never discussed it at all, refusing in this sense to become an object of his own speculations. Why otherwise did Heidegger remain silent on a subject addressed by so many other thinkers? If Heidegger knew that ongoing human activity requires "repression" of awareness of the extent to which we create ourselves and the categories by which we interpret reality, then he was in no position to pursue the meaning of the Holocaust to himself, in light of his choices, beliefs, and his stated need to remain rooted in German soil. That way lay the disintegration of values and beliefs, the disintegration of self, and madness.

of the social world as orderly and stable so they can view themselves as safely rooted in it.[25]

Heidegger had obviously mastered his personal struggle, distancing himself from the implications, generalizing his insight into a philosophical position that resonated especially among the young of that generation, those who had been in or near to the catastrophic war into which had been poured enormous amounts of energy and blood, and which, against all reason and logic as they saw it, they had lost. Heidegger's language of existence, anxiety, decision, nothingness, and death was appropriate for a generation compelled to deal with all of it for years (or had not dealt with it but wanted to, facing the world in either case as survivors), only to find themselves in a situation not of their making, giving evidence as well of being out of their control. Thus, Heidegger's insight into and elaboration of the ontological dilemma was crucial for the society; it was in fact a serious problem, if not originated in the war then certainly raised to a decisive level by it.*

Heidegger had hoped that Nazism would be "the instrument of fundamental and far-reaching changes in our German existence."[27] He was wrong, of course: He could not achieve any of his cultural or academic ambitions,[28] he could not prevent the disruption of academic life, let alone participate in the resurgence of society and the development of a "new man." Having joined

---

* This was a problem for Hitler, as I will explain in Chapter 3. However, it is useful to note here how much of what Hitler did and thought came out of or was crystallized by his experiences in the war, especially the constant need to face death, the constant fear of being overwhelmed by external forces. The whole concept of blitzkrieg, whatever its cognitive utility (maximum use of machines in a war of movement, a sensible technique for coping with refractory logistical problems), is based also on the model of traumatic experience, the sudden, overwhelming impact of events that cannot be mastered and that therefore paralyzes response. Hitler understood that this experience could be generalized as strategy, seizing upon the suggestion of rapid, deep armored penetration to disrupt command and supply structures, freezing the capacity to respond, fostering the impression of weakness and the hopelessness of resistance. Characteristically, Hitler told Rauschning that "the important thing [in war] is the sudden shock of an overwhelming fear of death."[26]

the Nazi party in May 1933, he refused to make himself responsible for all that the Nazis required of him, and he quit in February 1934. Official Nazi philosophers attacked him on the grounds that his commitment to the movement had been a masquerade from the start and that the movement could find its way out of the sterile pessimism suggested by his philosophy.[29]

It is important to stress, however, that Nazism had dignity for Heidegger. In the summer of 1935, he complained about the works then being peddled as National Socialism that actually had nothing to do "with the inner truth and greatness of this movement (namely the encounter between global technology and modern man)," a statement Heidegger decided to repeat in print in 1953, as he remained unconvinced that democracy was the political system most suitable for life in a technological world.[30] Heidegger had become disgusted with current practice; he had never intended to legitimate personal arrogance, bloated willfulness, or the amoral, violent domination of "inferior peoples." But Heidegger continued to see Germany playing a special role in the world, and he continued to see the moral force in National Socialism as a principle, a way of life, an idea that had been betrayed.

The pattern of expectation and disillusionment I have described occurred over and again with varying emphases and resolutions. Benn and Heidegger had great expectations for profound moral change; they were obviously morally distressed by the betrayal of their expectations.* Benn, who soon felt nothing but

---

* Ernst Jünger, an important figure in the postwar generation, is an interesting example of a militant and ferocious nationalist's moral recoil from Nazism. Jünger flirted with Nazism in the 1920s, because he thought that perhaps the Nazis had the strength and the will to mobilize Germany for the war he deemed both desirable and inevitable. But Jünger did not trust Hitler's decision to pursue a legal path to power; he thought that the Nazis had the capacity to betray the national revolution. Besides, he felt that he was more soldierly and aristocratic than the Nazis were. Although Goebbels had hopes for Jünger, as he did for the artistic community at large, once the Nazis had come to power, Jünger was lost to the Nazis and he made that quite clear to them. No one made more violent statements about war and life than Ernst Jünger, no one tried harder to reconcile people to the consequences of technology for the sake of preserving war as

hatred and contempt for the Nazis, their politics, their enforced silence, their impoverished conceptions of art and life, and, ultimately, their war, had sought refuge by 1936 in the army, the "aristocratic" form of internal emigration.[32] Heidegger was never as vigorous in his expression. He continued to teach and chose to remain in his native region because, he said, it was important for his work.

By contrast, Carl Schmitt, one of Germany's outstanding constitutional theorists, had opposed the Nazis before 1933 and was disappointed by Hitler's accession to power. Schmitt nevertheless turned around and worked hard to legitimate the regime. The Nazis, however, had as little patience with Schmitt's legal preoccupations as they had with Benn's conceptions of art; as far as Hitler was concerned, a conservative like Schmitt belonged by birth to a different species of man. The racists knew that Schmitt was not one of them, no matter how much he might pander to their interests publicly. Schmitt was attacked by the Nazis, particularly by the SS; but at a point when Benn and Heidegger had rejected the Nazi movement, Carl Schmitt continued to serve it, without any faith in its purposes and goals. Schmitt had joined the Party in 1933; his work for the Party ended in 1937. He was not disturbed, though he was distrusted on all sides.[33]

Schmitt's example was not inspiring, he did not set any standards for anyone. None of the three ever publicly repudiated his earlier position; all of them had joined as part of a social movement (indeed, Heidegger had invited Schmitt to join), but they quit as private individuals. All of them had served the Nazis to this extent at least. Still, the bases for making decisions were quite variable, the distinction between moral and instrumental decisions being an important and easily identifiable one.

Wilhelm Stapel, a prominent conservative Protestant theologian and polemicist, aggressively outspoken in his nationalism and anti-Semitism, was another who had no trouble justifying the

---

the highest sphere of manly endeavor. But Jünger was disgusted by Nazi violence and by what the Nazis finally made of their war. When Jünger's son died in the Italian campaign (1944), he said that he had joined the only true community of the war, the bereaved.[31]

shape of Hitler's rule as it initially emerged, especially as it was directed to the enhancement of German power. Stapel's Christianity was accommodated as readily to war, hierarchy, and submission to authority as it was to his nationalism; he urged the elevation of the volk against the so-called free-personality, the development of a national ethic consistent with "national biology,"[34] and the development of a new imperialism to establish the superiority of the German people and the hegemony of Germany on the continent.[35]

Stapel should have been a welcome ally of the Nazis, considering his glorification of the leadership principle, the volk, the nation, the military calling, and so forth.[36] But that proved not to be the case because Stapel's anti-Semitism was not racist enough. Nazi racists disapproved of Stapel's much-qualified condemnation of Jews, as they disapproved also of his divided loyalties: One could be Nazi or Christian, but not both. The Nazis made it abundantly clear, in any event, that Stapel's professional reflections on the problem of the Jews were not necessary and that he could abandon his role of "scholarly expert" in Hitler's state with no loss to them.[37]

Stapel's "inner protest" against the betrayal of his expectations was directed at first against the one-party state, the "tin-god Nazi party," a shameless rabble that took advantage of Hitler's weaknesses. If Hitler had had his way, things would have turned out differently. But then Stapel realized more and more, he said, that Hitler was himself driven by a demon, that in his megalomaniacal fury Hitler would guarantee the destruction of Germany. Stapel referred to the "abysmal vulgarity" of the Nazis, represented especially by Hitler, the denial of everything that National Socialism had once stood for. In the world created by the Nazis, "neither the old German world, nor the teutonic or northern one, but rather the world of the east and the south-east, victory belongs not to the noble, the good and the decent, but to the hard. Brutality and fanaticism, those loathesome characteristics of what is basest in mankind, are raised to virtues." By 1945, Stapel had concluded that a German victory under Hitler would have been a disaster, the Germans had been saved from a fate which Hitler's victory would have made inevitable, as there would have been no basis for recovery at all had Hitler prevailed. "It is enough to know that through temporary destruction God preserved us from eternal ruin." [38] *

Rudolf Binding, poet and novelist, who had won the Goethe Prize in 1932, responded to literary emigrants and foreign critics, as Benn had done, defending the regime against those who were repelled by it. In his reply to Romain Rolland's open letter of May 14, 1933, that is, after the book burning episode, Binding argued for this Germany, which he said was furiously desired by the whole population, regardless of the price. It is impossible for the world to overestimate, Binding insisted, the religious significance expressed in the depths of this revolution. Binding denied nothing—not the incitement to violence nor the racism that was bound to injure others, especially Jews, not the enforced exiles nor the assault on the life of ideas—nor did he seek to excuse the burning of the books. For all these things, as horrible as they seemed, were peripheral and did no harm to "the heart, the truth, of what is happening." But Binding, who died in 1938, nevertheless lived long enough to regret the corrupted spirit of Nazism. Binding, who both saw moral potential in Nazism and harbored doubts at the same time, was finally disillusioned by the degeneration of the society.[40]

Eduard Spranger, conservative philosopher and pedagogue, related at one point in October 1932 how he had attacked a

* Among Christian anti-Semites, the theologian Gerhard Kittel is very interesting, particularly by comparison with a man like Stapel. Kittel, who joined the Nazi party in May 1933, saw Nazism as a movement of popular renewal, resting on a Christian foundation. Kittel, a reputable scholar, considered himself a National Socialist theologian, dealing with "the Jewish problem" in religious, not racist terms. He remained in and worked for the Nazi party over the 12-year period of its rule, defending his work as serious research that had no harmful intent, to be distinguished from the vulgar anti-Semitism of Julius Streicher and Alfred Rosenberg. He maintained that his work was scientific, pursued for scholarly ends, that he had been compassionate to individual Jews, and that he had not contributed to or collaborated with genocidal policies.[39] Stapel, the deliberate anti-Semitic publicist, had come to see that Hitler was a disaster for Germany. Kittel, the dedicated scholar, professed to see nothing and, like Carl Schmitt and Gottfried Benn, considered that he had nothing to apologize for.

motion of censure against youthful Nazi student activists at a meeting of the Corporation of German Universities. But Spranger could not maintain the conviction that Nazi-inspired violent activism was a regrettable but transitory phase of a revolutionary period. Spranger could not conceive of a moral and intellectual life not guided by learning; there was a German university life he was prepared to defend. Spranger drew the line almost immediately. In February 1933, distressed, he said that the students who had accepted new responsibilities were beginning to resemble Metternich in their behavior toward opposition among the faculties or among other students. These aggressive youths were setting themselves up as arbitrary judges of writers and professors, always ready to condemn and to punish anyone who did not accept the principles of the national revolution. Spranger resigned his chair at the University of Berlin in protest.[41]

There was also the lamentable Gerhart Hauptmann, who initially reassured himself and everyone else that the fate of a few East European Jews was nothing to get excited about, only to howl in despair later (1938) that Hitler and his brown gang were determined to plunge the world into war; and Rudolf Pechel, editor of *Deutsche Rundschau*, who welcomed the Nazis as a bulwark against Bolshevism and ended up in Sachsenhausen concentration camp. In addition there were people such as Hans Carossa and Ina Seidel in literature, Ferdinand Sauerbruch in medicine, Ludwig Beck in the military, Wilhelm Niemöller in theology; in short, there were a large number of examples from all spheres of intellectual, academic, and professional endeavor characterized by similar experience.[42]

There is no point, however, in extending the list of individuals or examining in detail the variety of unique contents they expressed. I mean only to establish the heterogeneity of aspiration, striving, and insight, and the consistent sense among many highly educated Germans that Nazism might actually have contained the seeds of a genuine moral reawakening in order to raise certain questions: How could intellectually and culturally astute people, often quite experienced and well-intentioned from their own point of view, have seen moral potential in Nazism? And how could the stereotypical image of a peculiarly German predisposition to a unity of behavior based on psychological, familial, educational, and cultural impoverishment have continued to prevail when it is inadequate to a description of the range of people involved?

The Conservative Intelligentsia and the Common Sense World

The importance of these questions is underscored by reference to another writer, C. G. Jung, who was actually outside the German "system" as such. Jung was especially brash in the early days of Nazi rule, referring to the "formidable phenomenon of National Socialism," which was bound to shake "the Jewish categories of thought." Jung told his German audience that the "Aryan" unconscious contained the seeds of an unborn future, that the still youthful German people were fully capable of creating new cultural forms. The Jew, by contrast, could not contain in himself "the tensions of unborn futures." The Jew was rather something of a "nomad" who had never yet "created a cultural form of his own and as far as we can see never will, since all his instincts and talents require a more or less civilized nation to act as a host for [his] development."

Jung noted further that Jews share a peculiarity with women:

> Being physically weaker, they have to aim at the chinks in the armor of their adversary, and thanks to this technique, which has been forced on them through the centuries, the Jews themselves are best protected where others are most vulnerable.[43]

The Jewish categories of thought, however, should never have been applied to Germanic or Slavic Christendom: in particular, the "creative and intuitive depth of soul" of Germanic and Slavic Christendom should never have been explained in Freud's terms, for Freud could never have fathomed their mentality. Freud missed the "unparalleled tension and energy" in the German mind, which the Nazis had now succeeded in mobilizing as the world watched with astonished eyes, a result of his constant stress on "unrealizable infantile wishes and unresolved family resentments." Fanaticism is ever the brother of doubt, Jung said in reference to Freud's tenacious pursuit of his work; but Nazi fanaticism, deliberately encouraged as a means of treating the world, did not share this quality and was acceptable to Jung at this point because ("we") Aryans had been entrusted by fate with the task of creating a civilization. One-sided ideals, plans, and convictions were for that reason necessary.[44]

There was more, but the point is clear: [45] Jung fed a wishfulness that had already elevated instinct over reason, without defining any direction for it, that had elevated youth over age, German over Jew, and his judgment that Germans exhibited a youthfulness not yet fully weaned from barbarism was no insult to a group that reveled in barbarism. Jung claimed that he was not anti-Semitic ("I slipped up," he later told Leo Baeck [46]), and that is undoubtedly true, although he did little, especially in the beginning, to moderate live and extraordinary wishes and fears harbored by many Germans (e.g., the superiority of Aryans, the penetration and annihilation of healthy racial stock by alien, inferior elements).

Now we may leave to one side for a moment the violently aggressive conservative, nationalist, or racist spokesmen who insisted upon the desirability of war, the legitimacy of conquest, and the rest. Their notions about the world are quite different from those of Jung, who did not think about or interpret events in such terms. But how then did he interpret events, how did a man like Jung arrive at these conclusions?

Jung assimilated the Nazi experience to his own theoretical constructs, which, of course, he considered valid, making Nazism recognizable and predictable in these terms. Jung identified manifestations of the racial unconscious in archetypes and symbols, and accommodated his "interpretation" of Nazism to a common sense conception of how political leaders and movements behave and what they are likely to do.

It took Jung a while to get past his own ideological conception of the world and to subordinate his interests to moral concerns: Jung was too bourgeois to be able to justify Nazi transgressions forever. He had been taken in, primarily by his own thinking, which is not surprising. For people, no matter how intelligent and cultivated, can assimilate events, including the shocking and the bizarre, only in terms of expectations derived from experience and from their routine constructions of reality. The inability to assimilate events to expectations, or to the cognitive schemata that underlie interpretive processes, threatens the personal sense of stability, accounting for the tendency to force assimilation, to ignore or to "lose" information, and hence to misconstrue the import of events, a normal process of "regression in the service of the ego."

This process is not reflected on, it is not under conscious control, it is constant though people are not aware that it is happening; and it is facilitated by a variety of beliefs people hold, particularly that the personal and social worlds are more stable and orderly than they ever actually are, a belief undergirded by ideologically informed realistic and moral constructs (e.g., social "structure," defense "mechanisms"). Indeed, social order depends more on the need, and hence the willingness, to see the world as orderly than on any objective technique organized for the purpose.

Language is obviously the most immediate social technique for the organization of such thinking. But language is deceptive, as it does not have the fixity or stability of meaning implied by the dictionary definitions of the words that comprise it. The crucial feature of language is that meaning is not fixed—it is emergent—tied to specific situations and constantly changing. The meaning of language is really no more stable than the particular situations it may be used to describe. This flexibility derives from two related sources: First, every thought encompasses cognitive, moral, and wishful contents and is expressed at different levels simultaneously, implying the possibility for different meanings without necessarily implying different language; and the relative importance of these contents and levels is constantly shifting, in consistent ways according to location and circumstance, in unanticipated and unpredictable but systematic ways, as situations change for groups over time, and in idiosyncratic ways, as individuals may feel one way or another about things as they happen. Second, there is always a permissible latitude of interpretation of statements and events. There are limits, of course: Societies constrain the latitude of acceptable behavior, they maintain standards for defining anyone as sick, immoral, or criminal, and people routinely assume that shared behavior implies shared wishes, so that everyone appears to be acting from the same motives. Still, there is always room for a variety of legitimate alternative interpretations, and this is why we can never anticipate one single interpretation of an event by all the "normal" people who perceive it.

Therefore, it is very difficult to say what is meant by any but the simplest kind of statement, or how any complex statement is understood by others. Psychoanalytic practice has disclosed, if nothing else, just how complex the communicative process is, precisely because people can move so fast from one level of thought to another, effectively changing meaning without ever having changed the language. Moreover, social stability, or the stability of any institutional organization, depends upon the need people have not to ask what is meant by a statement, assuming always that people share a universe of discourse; and the distance established by not asking allows for the expression of subjectivity without requiring others to be entirely conscious of its import and having as a result to contend with it.

The complexity of this process is underscored when anyone deliberately seeks to mask meaning and has the skill and the determination to do it, a factor that has obvious significance for political communication, in which meaning is always masked, consciously or otherwise, and the desire to see through is inhibited. The attachment of any individual or group to social practice can be maintained as long as the social world is perceived as orderly, stable, and continuous. Political authority is perceived as the source of order, whether it represents traditional standards and expectations in familiar language or seeks to impose new standards and to change the language because the familiar has failed or appears to be failing. This is why political leaders, like dramatists, can manipulate the common capacity for a willing suspension of disbelief. But whereas people can follow the dramatist in his work because nothing is lost if they do, they must follow political leaders, because everything may be lost if they do not.

Thus, people see or hear things, but their understanding of them is consistent with their own sense of personal and social stability, as this is affected in any given instance by interest, morality, or wishfulness. Hence, Jung allowed himself to believe that he could protect or enhance his interests, serve the future of psychotherapy in Germany, and perhaps safely revenge himself against Freud, while describing some psychic reality—the truth value of which must override all considerations of vulgar chauvinist sentiment.* Jung lived in a world in which interest and wishfulness were constrained by morality, and he assumed that everyone else did too. He subsequently discovered that the cognitive and moral boundaries that affected his behavior did not necessarily affect others, and he was shocked when he realized the extent of his error.

The same, however, is essentially true of the German spokesmen for conservative and revolutionary nationalism I have reviewed in this section. These men were not afraid of civil strife and they glorified war, the necessity for which they justified on historical and moral grounds. No one could make more violent statements about war than the contemporary revolutionary nationalists (e.g., Ernst Jünger or Friedrich Hielscher). But they often came to despise Nazi arrogance and violence. These nationalists shared a language with the Nazis (will, discipline, heroic

---

* Konrad Lorenz may be viewed in the same terms. Lorenz wrote a paper as late as 1940 so loaded with Nazi terminology that even a sympathetic biographer could see "no scientifically respectable need for it to have been quite so clearly angled." Lorenz wanted to make a point to the Nazis about "domestication" (a human rather than a "racial" problem), and it is clear in context that he was not writing about Jews. Still, Lorenz managed to comment on "decadent art" as a sign of "symptoms of degeneracy," and on the phenomenon of domestication leading everywhere "to the fact that socially inferior human material is enabled . . . to penetrate and finally to annihilate the healthy nation. The selection for toughness, heroism, social utility . . . must be accomplished by some human institution. . . . The racial idea as the basis of our state has already accomplished much in this respect."[47] Lorenz claimed that he was not a racist or anti-Semitic and that his statement does not serve such purposes.

sacrifice, community, national power, the repudiation of abstract leveling constructs in favor of hierarchy, and instinct, understood as a natural or even primitive creative force). But they never meant the same thing by it.[48]

This language, and the realm of discourse that flowed from it,* had primarily moral implications for conservatives. That is, when Hitler repudiated the world of abstract legal concepts of right and wrong, of rationality, parliamentary procedure, and the like, these Germans especially moved into the moral world suggested by the romantic–heroic conceptions familiar to them. All the clamor for German science, art, music, leadership, religion, even when "instinct" was invoked, bolstered a sense of moral superiority that was confirmed by educators, doctors, church leaders, and other such authoritative figures. In this moral world the inevitability of struggle was taken for granted, but so too were the traditional constraints. However, this was not the world Hitler lived in, which was rather a world of grandiose entitlement,[49] unbounded by any of the moral scruples that bound the actions of others. Hitler did not recognize constraints, only positive or negative acts viewed from the standpoint of the racial community. This is why Hitler saw himself and his movement developing outside the mainstream of German history and tradition, as conventionally understood.

The real differences between these styles were masked in the first place by the use of language,[50] as traditional conservatives needed to assimilate evidence of violence and racism to their own version of reality, holding fast to the belief that their own moral commitments were universally shared, assuming a number of other things to be true as well: that the manifest violence was a

---

* The realm of discourse included, to a greater or lesser degree, the inevitability or desirability of struggle (war), a concept of hierarchy (as people and states are not equal and cannot live or be treated as if they are), the primacy of affect (as expressed in the binding power of the *Volksgemeinschaft,* not only different from but superior to the mechanical processes of Western liberal societies), and some version of the sacred quality of the past (expressed perhaps in terms of organic development over time or in terms of "eternal values"). I will return to the implications of this ideological network in Chapter 4.

transitional phenomenon and would subside; that Hitler's racial (and perhaps social) radicalism would be moderated once he was actually in power; that Hitler and the Nazis would be hemmed in by conservatives in government and the military, serving in the short run as a stabilizing influence but directed finally by more reliable people; and that the Nazis would fall from power when their policies were revealed as bankrupt, as surely they must.[51]

And the differences were masked by Hitler's skill and determination as a politician and orator. Hitler did not come to power on the strength of his anti-Semitism, after all. Hitler came to power rather on the strength of his appeal to traditional national and conservative sentiments, his promise of economic recovery and the end to class struggle, his promise to revoke the Versailles Treaty, thus restoring German power and independence, and his commitment to support and protect still-cherished beliefs and practices. In his first radio broadcast after coming to power, February 1, 1933, Hitler talked about Christianity as the foundation of national morality and the family as the basis of racial and political life. All through this period Hitler promised to eliminate civil strife as he reassured foreign powers of his benign intentions. But in fact Hitler had little patience with traditional conceptions of church and family; and on February 3, 1933, he had already told the military that Nazism would destroy Marxism root and branch and that Germany would seek relief from inevitable economic difficulties by conquest in the East.[52]

The Routine Interpretation of Reality in Common Sense Terms

THE FICTION OF CONTINUITY

People are not disposed to view the Germans' acceptance of Hitler generously; the commonly expressed feeling is "they should have known." Koppel Pinson once wrote that

> any informed German citizen who followed the antics of the Nazi formations . . . ; anyone who read the accounts of the behavior

of the Nazi faction in the Reichstag; anyone who read the telegram from Hitler to the Potempa killers and anyone who read the Boxheim documents—all these before 1933—was in a position to appraise exactly the true nature of the movement.[53]

Pinson's conclusion is understandable, but faulty. This is not how things work, from the standpoint of participants viewing their world prospectively, as bewildering, disruptive, and damaging events unfold before them. People assimilate information by invoking familiar patterns of thought, they strive as long as they can to construct a world that is predictable. Thus, Martin Buber concluded in February 1933 that

> the Hitler people will either remain in the government, regardless; then they'll be sent into battle against the proletariat which will split their party and will render them harmless. Or they'll quit [the government]; then there will be presumably a state of seige in which . . . the technical superiority of the army vis-à-vis the . . . [Nazis] will undoubtedly win the upper hand. As long as the present coalition prevails, any real persecution of Jews . . . or anti-Jewish legislation is unthinkable.

Buber apparently thought that legislation against the Jews would become possible only if the given constellation of power in February 1933 shifted in favor of the Nazis, which he considered unlikely.[54] The extraordinary capacity for violence that Pinson retrospectively referred to did not suggest extraordinary consequences to Buber at the time. It was still a common sense political world in his mind.

The prominent physicist, Leo Szilard, described a conversation he had with his friend Michael Polanyi (the director of a division of the Kaiser Wilhelm Institute for Physical Chemistry, and the brother of Karl Polanyi) in that very same period, after Hitler had come to power. Szilard himself "had no doubt what would happen," and he lived with his bags packed. But Polanyi

> like many other people, took a very optimistic view of the situation. They all thought that civilized Germans would not stand for anything really rough happening. The reason that I took the opposite position was based on observation of small and insignificant things.

Szilard did not like the instrumental view Germans had of the world; he had realized by 1931, he said, that Hitler would win, not because the Nazis were so strong but because others would not resist if they saw no advantage in it.

Although it would have been difficult for him, Polanyi could readily have left Germany at that point; unlike most other people, he had a job waiting for him in England. But Polanyi did not want to leave his laboratory, and he decided at first to wait the Nazis out. Szilard's sense of danger increased with the Reichstag fire, which he blamed on the Nazis. Polanyi was incredulous: He could not believe that the government had a hand in it. Szilard left Germany and so, of course, did Polanyi, shortly thereafter. But the point is that neither of these two men took into account what Pinson was convinced any reasonable person should have realized—and they could hardly be considered uninformed men.[55] There were many other such instances as well. Max Planck urged caution in treating with the regime, assuming the violence to be part of a transitional period. Planck said as much to a frightened colleague who wanted to leave: The present situation would blow over and things would return to normal.[56]

Thus, even individuals as brilliant as Buber, Polanyi, Planck, and Jung interpreted the contemporary reality in rather common sense terms bound by idiosyncratic expressions of interest, morality, and wishfulness. Their brilliance was exhibited in their spheres of competence, as people strive for objectivity and insight in areas of professional competence. Otherwise, outstanding individuals treat the world like more ordinary people, interpreting events by means of "situational guesses," without the means for getting at the truth, and perhaps, in any given instance, without the desire. People accommodate events to their needs by common sense explanations in language made available by the culture. Szilard feared the Nazis, to be sure, but his understanding of the situation was not dazzling and he was not bragging: When he said he saw "small and insignificant things," he meant that he had a hunch things would not turn out well, and that is all.

The social world is interpreted this way because it is at risk for everyone at all times, rendering personal stability problematic. Action is possible only in the context of a coherent, comprehen-

understanding of the world that at the same time defines one's activity as continuous within it. In fact, things had changed radically for Germany when Hitler came to power—and some Germans, at any rate, took Nazism to be a liberating force that could free them from the staid, failed, traditional conventions, promising them a life of adventure, excitement, and violence, which they welcomed.

But most Germans, especially middle class Germans at all levels, still cherished the traditional conventions, and they wanted what Hitler had promised: a strong Germany, the reversal of Versailles, an end to civil strife and class conflict, and support for family and religion. Most Germans just wanted the crisis resolved, and they took Hitler at his word: Standing above interests and parties, he would serve all of them alike. Thus, people saw the violence and they heard the threats, but they wanted to treat these things as isolated occurrences and not as part of a pattern, particularly if these occurrences did not threaten them. Ordinary German citizens had to make sense of the world they lived in, in terms they were familiar with, and so they told themselves that one cannot make an omelette without breaking eggs, one cannot clean house without chipping some porcelain, one cannot plane wood without shavings, the sickest cases get the strongest medicine, the devil is driven out by Beelzebub, and so on. It was mostly a matter of common sense: What was Hitler going to do anyway, kill all the Jews? [57]

The process by which people accommodated themselves to Nazism is clarified whenever we find a day-to-day description of how people lived through events. It is possible to see how people convinced themselves that things would turn out all right, how they were manipulated, their goodwill exploited, how personal pressures deflected attention and undermined resistance, how a highly skilled and utterly determined politican encouraged belief in a familiar world by appeals to interest and morality, fostering a sense of continuity with the past. Even when one or another individual began to realize that the world was becoming unfamiliar and even dangerous, he still allowed himself reason to hope.[58] The situation was such that even many Jews could continue to hope or to assume that they would be allowed to participate in the new

Reich, or that Hitler was an honorable man who would treat responsible German citizens decently.[59]

What had it been like in retrospect for an individual who had thrown himself into the Nazi movement?

> I lived with great intensity through the late twenties and early thirties: the collapse of Weimar democracy, party hatred, sharpening of class distinctions, economic disaster, six million unemployed and the despair of any hope in the future. Then, on the threshold of my career, though conscious of the brutal power and oppressiveness, there was also a faith—a faith later terribly betrayed and sometimes self-betraying—in possibly a new beginning, in a future strengthened by common purpose and service to the whole. So for me it was not simply a question of opportunism, of choosing the easy way, but rather of deciding whether the things that shocked me were a challenge to self-isolating resistance or whether, on the other hand, my positive reactions were a call to active cooperation. My background and education—I come from the eastern frontier district and grew up in unquestioning acceptance of what before the First World War was known as a form of nationalism—led me to cooperate. It is true that I soon discovered with increasing conviction that the primitive, simplistic, anti-intellectual biologically-collective and brutal aspects of Nazism were totally unacceptable to the individualist and humanitarian tradition to which I belonged. But mindful of comparative developments in Russia, Italy and the Iberian peninsula, it seemed clear to me that the age of humanism, individualism and historically derived culture was being superseded by an era of new forms and ideals which appeared barbaric to those rooted in tradition, but of which the Nazi version seemed to me preferable to the Bolshevik one. In view of the horrors of the French Revolution and of Stalin's mass murders, I took the early appearance of the so-repellent Nazi violence for an oppressive but apparently inevitable accompaniment of revolutionary breakthrough. The basic conflict in which I was so deeply involved lay in the discrepancy between my conviction of the irresistible power of a collectively-directed mass society with its abandonment of former cultural and scholarly traditions, and my unshakeable certainty that for the rest of my life I would remain loyal to cultural, scholarly and esthetic values and standards which were doomed to disappear.[60]

The expressed sense of confusion resolved by a decision to participate, subsequently regretted, is believable enough—though the statement is heavily rationalized and worthy of an academic. On the other hand, how did people respond prospectively to events, particularly people who had every reason to fear and despise Nazism? Charlotte E. Zernik described her father's attempt to calm his family when they learned that Hitler had become chancellor: "It won't really be so bad," he said; "even [Hitler] can't make bricks without straw." [61] These words, Zernik wrote, are still ringing in her ears. Victor Klemperer, who survived the war in Germany though he was Jewish, was assured in early 1933 that "the fuss about the Jews is only propaganda. When Hitler's running things, you'll see, he'll have better things to do than worry about Jews." [62] James Frank, who resigned his chair at Göttingen University in protest against Nazi treatment of Jewish scientists and educators, was admonished for his hastiness by a colleague who added, "Nothing is eaten as hot as it's cooked." [63] There were any number of such statements, meant to keep the common sense view of the world intact, to indicate that the situation was bad but only temporarily so, that it was bound to turn in a favorable, that is, a familiar direction.* Joel König wrote about his father's ability to remain optimistic in the face of events—Rabbi König thought that the rise of the Nazis must deal a severe blow to anti-Semitism because responsible people would not pursue such a line in company with that gang of murderers. There was no reason to despair because the Jews were not alone: "There are still plenty of Germans who will have nothing to do with the Nazis." [64]

The responses of such individuals to the events that preceded and followed the Nazi seizure of power must therefore be viewed as situational and emergent, continuous with the past in their

* Wilhelm Hoegner, an official of the Social Democratic Party, relates how his young daughter responded to a classmate's threatening comments about her father's future, after the Nazis had come to power: "Apparently my daughter replied, 'The world is round and is constantly turning. What's down today can be up tomorrow.'" [65] Perhaps it is inappropriate to cite a child's response in this context, as children are likely to say such things in any case. But she does not seem to have been any more or less naive than many of her adult contemporaries, including the leadership of the Social Democratic Party who, according to Hoegner, completely misperceived the nature of the danger facing them.

own minds, as they were compelled to construct the world to make it appear so. Their behavior was not the result of predispositions systematically derived from socially imposed forms of psychic deterioration. There is no theoretical or empirical advantage to this viewpoint—with respect to these people or to any element of the German population. It forces the observer to think in terms of continuity regardless of what anyone actually said or did, and it ignores the confusing, rapidly shifting, contingent reality in the period 1930–1933 and the misconceptions that attended and fostered the Nazi triumph, as it belies the Nazis' own fears after the elections of November 1932 that having lost ground for the first time, exhausted and financially strapped as a result of their enormous efforts in 1932, they had passed their peak and were faced with decline. Clearly, the Nazis had not just gone from strength to strength; there had been a dramatic shift in voter interest over a short period of time in the winter of 1932, and the movement was vulnerable to such short-run shifts that cannot be explained in theory but that were certainly sufficient to cause alarm and anxiety among the leadership and in the movement at large. The population that provided the Nazis with considerable support was not altogether a stable one, and the Nazis' assumption of power depended finally more on political manipulations than on popular success. It cannot be emphasized too strongly that the Nazis never won a majority in any election through March 1933, when Hitler had already become chancellor of the Reich and the Nazis could bring the force of government to bear against their opponents.[66]

The "authoritarian personality" concept, or any concept that implies a unified network of motives and perceptions based on psychological deterioration before the events in question, is therefore inappropriate and misleading. The many people whose decisions and perceptions have been reviewed here cannot be described as lacking in insight or incapable of discriminating judgments, except as they were perhaps affected by an immediate situation. Werner Forssmann, Nobel laureate in medicine and an outstanding surgeon, became a Nazi, he said, because of professional frustration engendered by a conservative medical establishment.[67] It would be hard to argue that such a man, or men like Heidegger and Benn, had only a few limited psychological techniques at their disposal or that they were driven to join an authoritarian movement because they were too psychologically deteriorated to resist.

In fact, there is no way to accommodate the people involved to one characterological conception, or even a few. They were too heterogeneous in motive and perception, and nowhere is this heterogeneity more immediately evident than in psychoanalytic studies of elite Nazis[68] and in sociological studies that made a deliberate effort to acquire subjective data, such as the autobiographical statements solicited and collected by Theodore Abel in the 1930s.[69] *

Henry V. Dicks, who has probably had more personal and professional experience with this problem than anyone else, noted in his analysis of 1000 German prisoners of war, interviewed between 1942 and 1944, that 11% could be classified as fanatical Nazis and another 25% as "believers with reservations" (actually people who tended to be conservative and nationalist in a more traditional German sense). The remaining 64% included the indifferent ("We've had the Kaiser, Hindenburg, and now Hitler, but life must go on, etc."), 40%; the divided or passive anti-Nazis, 15%; and active, democratic anti-Nazis, 9%. Moreover, those classified as fanatically Nazi were not dynamically homogeneous among themselves, let alone in relation to the others. Finally,

> the results of comparing my personal ratings of those who scored low on Nazi fanaticism . . . [show] a spread of the variables along a range, rather than a polarization into black and white, even on

---

* Each individual, no doubt, had reasons for joining (or quitting) the movement, and these reasons are traceable to idiosyncratic, unconscious mental activity. But this is impossible to determine in specific instances except as biographical information might be available. However, the diversity of behavior and language, indicative of competitive or defiant strivings, or of longings for nurturance and protection, the expression of compulsive, "masculine" activity, noisy, provocative demonstrations, violent repudiations of moral sentiment, aggrandized self-regard, sentimental idealizations of comradeship, of innate superiority, or of opportunism and indifference, and expressed opposition to all of this from people of similar backgrounds does not suggest psychic homogeneity or the existence of a German "type."

those variables that turned out to be discriminating between the personalities of those rated High and those rated Low.

The tendencies Dicks found to be consistent among high scorers matched the findings of the independently achieved Authoritarian Personality studies. But the authors of those studies had also concluded that Nazis tended to be heterogeneous in character.[70]

Some Critical Comments on Social Scientific Views of Nazism

The principal theoretical ideas about Nazism are distorting because they imply that the personal and social worlds are more orderly and stable than they ever actually are.* The idea of continuity may be reassuring to observers, but it is irrelevant to the experiences of large numbers of Germans who were caught in a crisis situation that rendered routine and valued modes of thought dysfunctional or inapplicable from the standpoint of their conscious expectations. The idea of homogeneity of character may be similarly reassuring, and it is similarly irrelevant. Sociologists and historians were bound to underestimate the effects of the immediate situation, to fail to account adequately for the feelings that arose in the face of actual or threatened discontinuities of experience, and therefore to misconstrue the subjective perceptions and motivations that prompted behavior at the time.

These ideas are not confirmed by the evidence, though it is not hard to see why, from a theoretical point of view, they retain their force. Without some conception of personal and social continuity (especially predispositions to behavior resulting from class and family background), and without some technique like ideal–typical description of character, there would seem no longer to be

* The ideas are not much different from the most universally held common sense ideas about the world, and social scientific conclusions about Nazism—that it appealed to the middle and lower-middle classes and to the young—do not go much beyond the common sense observations of contemporaries.

a common cause for a common effect, and the ability to think systematically, in the sense of measuring and counting, would be frustrated.

However, measuring and counting are not always the best way to analyze events, and they are particularly inappropriate to the analysis of Nazism, which requires insight into subjective processes. At best, ideas of continuity and homogeneity are useful in the short run because they allow for the study of objective categories (class, age, occupation, region, religion) in numerate terms. At worst, they are an evasion, masking an ideologically informed conception of the uses of social science, serving, paradoxically, to direct attention away from the crucial problem of human subjectivity.

It must be recalled that this strategy of investigation, the linking of class, family, and unity of character, was developed by cognitively disciplined Marxist theorists who, armed in addition with psychoanalysis, were determined to defend the morality of rationality and democratic participation against an enemy determined to destroy it. These theorists elaborated an ideological position that served to create a single enemy out of a variety of right-wing tendencies and was then amplified by liberal and Marxist theorists to make the broader case against "totalitarianism."[71] The ideological purpose sustains the strategy because it promises an objective, "scientific," even numerate solution to a complex problem involving human subjectivity, rendering the observer's world coherent and helping liberal, inclusive, and cognitively disciplined societies to struggle against such enemies, in terms they can accept. It is not that the strategy as such is cognitively useless, although it is always heavily criticized on empirical grounds; it is rather that cognitive utility was never its primary function. To define anyone who saw any potential moral worth in Nazism in 1933 as predisposed to authoritarian behavior, the result of a process of psychic deterioration, is to say that all orientations to the world that did not include rationality and democratic participation as bases for action were irrational—and hence that all German conservatism was irrational, that there were structural and dynamic similarities between Nazis and conservatives because they all shared a language of commitment to affect, hierarchy, authority, and struggle, and they all "really" knew what Hitler was about.

This ideological construct has resulted in a lot of confusion and has consistently interfered with the analysis and interpretation of Nazism and particularly of the rise of Nazism to power. German conservatives were not all alike, and there were many bases for the acceptance of Nazism when it first gathered strength. As Carl Zuckmayer wrote:

> It would be unfair altogether, altogether wrong and misguided, to issue a wholesale condemnation of the large numbers of Germans who poured into the Nazi movement at the beginning of the thirties. For at this time the hopeless were joined by the hopeful, the idealists, the believers who in their wishful thinking imagined that something ethical and decent underlay . . . this movement.

And as one German conservative later wrote.

> The demonic traits of the system were then still hidden from me, or were subordinated to other traits which even a non-National Socialist could welcome, simply as a fellow German. The real spirit of the Third Reich manifested itself to us only gradually.[72]

Such an ideological construct may be necessary for the sake of struggle, but if it is employed without reflection, then it inhibits insight, not only into the motives of those who participated in these events but also into the social and psychological bases of rationality and participation as a *moral* orientation itself.[73] From the standpoint of traditional German conservatives, for example, as they faced an immediate crisis and interpreted the potentialities of Nazism, the problem simply could not be viewed in terms of Western and Marxist rationality and German irrationality. Rather, from their own standpoint, there were two moralities involved, one focused on cognitive–rational orientations and one focused on affective–experiential orientations. These two moralities could not be hierarchically evaluated on principle. Either of the two could serve the integrative purposes of social organization, just as either could serve disintegrative, destructive purposes: People have just as readily been murdered in the name of reason and justice as in the name of order and authority. People like Heidegger, Benn, Binding, and others saw moral potential in a system of national socialism based on order and authority in these terms, some of them continuing to see it even after they had been embittered by Nazi practice, convinced that a worthwhile movement had been taken over by a pack of cutthroats.[74]

There may have been pathological strivings among German conservatives and other adherents of Nazism; however, this must be established, not assumed, and any interpretation must take into account the different subjectivities of the people involved, in terms of the effects of immediate circumstances on cognitive and affective processes. A comprehensive, empirically adequate interpretation of the appeal of Nazism must account for the ways in which people assimilated and controlled threatened or actual discontinuities of experience and reconciled in their own minds unanticipated discrepancies betwen everyday expectations and the ability to realize them. Such an interpretation must account for the well-established objective data, particularly the heterogeneous composition of the Nazi movement, and for the various subjective strivings of the people involved, particularly from the time Nazism emerged as a successful mass movement to the Holocaust.

This requires in the first place a sociological standpoint that can integrate psychoanalytic insight on some basis other than impulse, transgression, and punishment. The traditional psychoanalytic orientation to drives and the control of drive expression through familial interventions is inappropriate to an explanation of the appeal of Nazism. That appeal followed from the threatened or actual inability of large numbers of people to act on socially valued character traits, skills, and ideals, and from the ability of Hitler and the Nazi movement to appear as if they were capable of turning a disrupted and potentially chaotic situation around.

Notes

1. This dispute, and Hitler's resolution of it, is ably described by Hildegard Brenner, "Art in the Political Power Struggle, 1933–34," in Hajo Holborn, ed., *Republic to Reich: The Making of the Nazi Revolu-*

tion (New York, 1972), pp. 395–432. (I have employed throughout an anglicized version of "Völkisch," as a matter of convenience.)

2. Gottfried Benn, *Gesammelte Werke*, ed. Dieter Wellershoff, 4 vols. (Wiesbaden, 1959–1961), "Kunst und Macht," 4, p. 396. Hereafter referred to by author, title of piece, volume, and page.

3. Benn, "Answer to the Literary Emigrants," in E. B. Ashton, ed., *Primal Vision: Selected Writings* (London, 1971), p. 48.

4. Benn, "Der neue Staat und die Intellektuellen," 1, pp. 443, 448.

5. For a survey of Benn's writings in 1933–1934, see Josef Wulf, *Literatur und Dichtung im Dritten Reich: Eine Dokumentation* (Gütersloh, 1963), pp. 113–115. On Benn's defense of eugenic policy, see Benn, "Züchtung I," 1, pp. 219–220.

6. Klaus Mann had invited Benn to join the literary community in exile, claiming that there would be no one left in Germany who could understand his art and that in Western experience the celebration of the irrational always culminated in barbarism. Benn responded in the terms just noted. Klaus Mann's letter and Benn's response, "Answer to the Literary Emigrants," are both in Ashton, ed., *Primal Vision*, pp. xiv–xv, 46–53; see also Benn's retrospective comments in his "Doppelleben," 4, pp. 70–74.

7. Benn, "The New Literary Season," in Ashton, ed., *Primal Vision*, p. 42.

8. Benn, "Der neue Staat und die Intellektuellen," 1, p. 447; "Answer to the Literary Emigrants," in Ashton, ed., *Primal Vision*, pp. 51–52.

9. Benn, "Rede auf Stefan George," 1, pp. 473, 474–476; "Rede auf Marinetti," 1, p. 481; "Probleme der Lyrik," 1, pp. 524, 507–508; "Expressionismus," 1, p. 252.

10. Benn, "Doppelleben," 4, p. 136. The work of art should be historically ineffective. Style is superior to truth because it carries the proof of existence in itself. Benn abhorred didactic art and he denied the social function of art as a symbolic process that organizes meaning for people, something peculiarly gifted people do for others. Benn's position would be plausible if form did not harbor a message of its own, particularly when it is so busily rationalized and when it is clearly understood at the same time that the unity of the personality is a problematic thing, that the "loss of the center" is an abiding human fear. The discussion of form for Benn played the same role as the discussion of being did for Heidegger. It was a way of controlling a sense of chaos; the discussion of form was as much derived from necessity as any content. Benn must have understood this and it would account for his pessimism.

11. Benn, "Rede auf Stefan George," 1, pp. 476–77, 473.

12. Benn, "Expressionismus," 1, p. 240. Josef Wulf, *Die bildenden Künste im Dritten Reich: Eine Dokumentation* (Gütersloh, 1963), pp. 44, 50. National Socialism was treated as a positive organization of life that could release artistic energy. *Ibid.*, pp. 118–123.

13. Benn, "Doppelleben," 4, p. 128.

14. Benn, "Expressionismus," 1, pp. 241, 248–250; Benn, *Ausgewählte Briefe* (Wiesbaden, 1957), letter to Ewald Wasmuth, October 18, 1936, p. 74.

15. Benn, "Lebensweg eines Intellektualisten," 4, p. 20–23.

16. Benn, *Ausgewählte Briefe*, letters to Ina Seidel of August 27, 1934, and December 12, 1934, pp. 58, 62; to Thea Sternheim, August 12, 1949, p. 169; "Doppelleben," 4, pp. 94, 110. Benn repudiated his maxim that all is permissible that leads to experience. Such a conception was appropriate for gentlemen, but the Nazis were not gentlemen.

17. Hence, for example, K. D. Bracher's observation of innumerable instances of voluntary "coordination" among academics and his contention that by 1939 the administrators of higher education in Germany had gone a long way towards meeting the demands of the Nazi leadership. *The German Dictatorship*, trans. Jean Steinberg (New York, 1971), pp. 266–272.

18. Gottfried Benn pointed out that not everyone who supported the movement was opportunistic—referring to himself, first of all. "Doppelleben," 4, p. 89. On the other hand, almost no one in the churches in 1933 publicly criticized the abolition of civil and political liberties; churchmen tended rather to applaud that step as heralding "the establishment of order," as the Nazis destroyed Marxism in Germany and seemed capable as well of reversing the Versailles Treaty. See Dietrich Bronder, *Bevor Hitler kam* (Hannover, 1964), pp. 276, 171, 173, on the Lutheran Bishop Otto Dibelius: "One cannot fail to appreciate that Jewry plays a leading role among all the disruptive phenomena of modern civilization." The Nobel laureates, Philip Lenard and Johannes Stark, and long-time Nazi professors Ernst Krieck and Alfred Baeumler were rather idiosyncratic Nazis. On Lenard and Stark see Alan D. Beyerchen, *Scientists under Hitler: Politics and the Physics Community in the Third Reich* (New Haven, Conn., 1977), pp. 79–131; on Krieck and Baeumler, see Jean-Michel Palmier, *Les Écrits Politiques de Heidegger* (Paris, 1968), pp. 93–97, 299–331.

19. See Bronder, *Bevor Hitler kam*, pp. 14–15.

20. Guido Schneeberger, *Nachlese zu Heidegger: Dokumente zu seinem Leben und Denken* (Bern: 1962), pp. 144, 136; Palmier, *Les Écrits Politiques de Heidegger*, p. 290. Heidegger was named rector of Freiburg University in April 1933, and his rector's address, which has become notorious, justified the elimination of the contemporary standards of academic freedom which Heidegger said were false in any case and merely signified a license to act or not as one saw fit: The concept of freedom for the German student under the guidance of Hitler would now be led back to its truth. Heidegger joined the Nazi Party in May 1933 and later was also the most prominent academic to lend his name to a "Profession of Faith" signed by 960 members of the academic community on the occasion of Germany's withdrawal from the League of Nations. In his statement, Heidegger was anxious to deny that this step portended a return to barbarism, an irruption of lawlessness, the denial of the creativity of a cultured people, or the destruction of

its historical traditions. On the contrary, what it portended he said was "a resurgence of youth rooted in the past and purified." Schneeberger, *Nachlese zu Heidegger*, pp. 91–93, 149; Palmier, *Les Écrits Politiques de Heidegger*, pp. 78–88, 114–116, 288.

21. Hannah Arendt, "Martin Heidegger at Eighty," *The New York Review* of Books, vol. 17, no. 6 (October 21, 1971), 50–54.

22. Freud and other psychoanalysts were aware of the potentially traumatic impact of external events overwhelming in their intensity, as evidenced by the "war neuroses." Following some suggestions of Ferenczi's, Freud did address himself to the threat to narcissistic integrity that follows specifically from conflicts between ego and reality involving primarily not what we do (actually or wishfully), but what we are. The issue was not amplified then, but it certainly has been since, as narcissism has become the primary focus of contemporary psychoanalysis. See Meyer S. Gunther, "Freud as Expert Witness: Wagner Jauregg and the Problem of the War Neuroses," in John E. Gedo, *et al.*, eds., *The Annual of Psychoanalysis*, vol. 2 (New York: 1975), pp. 10–23.

23. Richard D. Chessick, "Defective Ego Feeling and the Quest for Being in the Borderline Patient," *International Journal of Psychoanalytic Psychotherapy*, vol. 3 (1974): 73–89.

24. Heidegger saw Europe, and Germany in particular, caught in a pincers between the Soviet Union and the United States, two societies metaphysically alike in their leveling tendencies and their devotion to number, destroying every world-creating impulse of the spirit, standing for the same dreary technological frenzy, the same unrestricted organization of the average man. In 1935 Heidegger still saw "new spiritual energies unfolding historically from out of the center," that is, Germany. Martin Heidegger, *An Introduction to Metaphysics*, trans. Ralph Manheim (New Haven, Conn., 1959), pp. 37, 39, 45. Western scholars have viewed Nazi Germany and the Soviet Union as two versions of the same totalitarian orientation; the Soviets of course officially view Nazi Germany and the United States as two versions of the same imperialist orientation; and Heidegger here expresses a German conservative view of the Soviet Union and the United States as two versions of the same technological–rational and inclusive orientation. In particular, service and sacrifice in the absence of class barriers and rank, and voluntary labor for the community were important to Heidegger. See Palmier *Les Écrits Politiques de Heidegger*, pp. 120–123, 142–145, 224–225.

25. George S. Klein, *Psychoanalytic Theory* (New York, 1976), pp. 133–135.

26. Hermann Rauschning, *Hitler Speaks* (London, 1939), p. 90.

27. Schneeberger, *Nachlese zu Heidegger*, pp. 149–150. Heidegger explained later that he had believed in Hitler and Nazism—though he was to be bitterly disappointed—for a variety of practical and moral reasons: He thought he could help effect necessary reforms in the university structure

and reduce class barriers, making the schools more accessible to the poor while keeping the university open to different currents of opinion. But the most important thing was the mobilization of the academic community to help realize the moral promise, to develop the spiritual and intellectual resources of the movement. For with the end of Weimar the Germans had elected to turn away from a barren and impotent mode of thought, as the National Socialist revolution was not simply "the seizure of an existing power in the state by another suitably qualified party." It was, as noted, "the instrument of fundamental and far-reaching changes in our German existence."

28. Palmier, *Les Écrits Politiques de Heidegger*, pp. 73-74.

29. *Ibid.*, pp. 93-99, 280, 282.

30. Heidegger, *An Introduction to Metaphysics*, p. 199; Palmier, *Les Écrits Politiques de Heidegger*, pp. 215, 268. On the comments made in 1935 and confirmed in 1953, *ibid.*, pp. 278-279. The emphasis on technology is discussed by Palmier in relation to Ernst Jünger, see n. 31, immediately following. Heidegger's statement referred to national socialism, not National Socialism, and that was all right in his own mind.

31. Ernst Jünger, like many who had espoused fascism in some form in the interwar period, had been driven by his experiences in the Great War, particularly by the decisive establishment of the primacy of technology, to a narrow, brutal either–or: One could either ignore the harsh reality and be destroyed by it, or one could master the reality by identifying with the technology, sharing thereby in its awesome power. In a world dominated by machines, there were only victims or executioners, and Jünger chose to be an executioner.

Jünger held that Germany had been spiritually unprepared for the kind of total mobilization ultimately required by World War I, from which something entirely unexpected had emerged: The massive, stunning fire power had shattered the romantic world with its exaltation of the chivalric–aristocratic warrior. Ernst Jünger, "Die totale Mobilmachung," in Ernst Jünger, ed., *Krieg und Krieger* (Berlin, 1930), pp. 11, 14. As a British journalist at the battle of the Somme wrote, "Courage isn't what it used to be. The machine runs over us and we can't stop it." Quoted in Christopher Martin, *The Battle of the Somme* (London, 1973), pp. 27-28. Jünger, who had shared that romantic spirit, recovered himself, reveling in its destruction, which cleared the way for a ferocious, masculine activity. Palmier, *Les Écrits Politiques de Heidegger*, p. 189. The war had turned out to be the prelude to a new era, and Jünger urged the development of hard, unintellectual men who would encompass in themselves cold technique and fiery passion, and who would therefore be able to rule themselves and everyone else according to modern requirements. Hans-Peter Schwarz, *Der konservative Anarchist: Politik und Zeitkritik Ernst Jüngers* (Freiburg im Breisgau,

1962), p. 69; Ernst Jünger, *Der Arbeiter: Herrschaft und Gestalt* (Hamburg, 1932), p. 34.

Jünger concluded that because war was desirable and inevitable it was important to promote the mechanization of war and of life itself. The autonomous individual, rendered obsolete by history, must be superseded by a technologically proficient worker type for whom the uniform would be the standard dress, the visible expression of the necessary subordination of the individual to the whole and of the willingness to accept war as a form of work, the trench, like the factory, as a zone of work, and the soldier as a worker in the sphere of death. Jünger, *Der Arbeiter*, pp. 104–105, 120, 134, 1443–144; Jünger, "Die totale Mobilmachung," p. 15; Schwarz, *Der konservative Anarchist*, pp. 60, 69–70; Palmier, *Les Écrits Politiques de Heidegger*, pp. 167–215; and Walter Struve, *Elites against Democracy: Leadership Ideals in Bourgeois Political Thought in Germany, 1890–1933*, (Princeton, N.J., 1973), pp. 377–414.

Jünger celebrated the fall of bourgeois society, thrown over by the unanticipated incursion of elemental forces that destroyed the cult of reason that had animated bourgeois life. Jünger, *Der Arbeiter*, p. 47; Schwarz, *Der konservative Anarchist*, pp. 71–72. But all the other traditional forms of authority had fallen as well. There was no possibility of restoring the monarchy or aristocratic hierarchy. All other things aside, action was now of such scope and required such brutality that it could only be unfolded in the name of the community. Besides, once catastrophe had emerged as the a priori of a new order of life and the world seemed about to be transformed into a landscape of factories and combat zones, the past no longer mattered.

Jünger believed in the cult of war. He tried to accommodate bourgeois technology to the soldierly spirit, to synthesize the old and the new, to save war as the highest sphere of manly endeavor. This new synthesis was necessary because although the bourgeoisie had changed the world with their machines they could not be allowed to impose their sterile way of life. The bourgeoisie certainly had no cause to complain about such a turn of events, having already required duty, discipline, and communal effort of everyone, merely for the sake of private profit; and if men could be mobilized for the sake of private profit, then why not for the sake of national power. Schwarz, *Der konservative Anarchist*, p. 85.

For a while, in the 1920s, Jünger was close to Hitler and the Nazis because he thought that perhaps they had the power to achieve this kind of radical ambition, to restore German power for the sake of resuming the combat temporarily interrupted at an unfavorable moment in 1918. Schwarz, *Des konservative Anarchist*, pp. 117–120. Still Jünger remained aloof from the movement. The Nazis worried too much about party organization and legal paths to power for his taste. They were apparently willing to work within the hated Weimar system, and they seemed, therefore, to harbor the potential for betraying Jünger's ideals.

The differences between Jünger and the Nazis may be viewed as political, social, or intellectual—but fundamentally the differences were moral. For despite all the bluster Jünger was and remained an elitist, aristocratic, chivalric soldier, wedded to a warrior code that was as remote in its way from Nazi practice as anything could be. It was part of the code to be merciful and to have compassion for the unfortunate. To the Nazis, especially the racial Nazis, this was hopeless sentimentality. It was part of the code never to become a mere instrument, for the soldier is a man of conscience. By contrast, Hitler was interested in men only as instruments and professedly he was not a man of conscience. Jünger, moreover, could not bring himself to hate his warrior enemy, because he had too much respect for the front-line soldiers on the other side who shared the code and were a part of the chivalric order. Jünger tried to assimilate the meaning of the new technological reality, especially mechanized warfare, in the manner suggested. But he was also compelled to find a way out, to work back to his earlier heroic idealism, his spirit as a young officer in the Great War. Jünger was compelled to do so by the Nazis themselves, as they revealed what Jünger considered to be their great capacity for barbarism. Schwarz, *Der konservative Anarchist*, pp. 59, 118. Later he claimed that the vision of *Der Arbeiter* was really a forecast and not a prescription. Struve, *Elites Against Democracy*, p. 384. I should note that as far as Goebbels was concerned, both Jünger and Benn were lost to Nazism by 1934—and Jünger had certainly made his feelings clear to the regime by then. See Hildegard Brenner, *Die Kunstpolitik des Nationalsozialismus* (Hamburg, 1963), p. 117; Wulf, *Literatur und Dichtung*, pp. 11, 35.

Jünger's ideal was a warrior brotherhood; he hated democracy in all its forms. But still his sentiments had nothing in common with Hitler's conception of the racial struggle for existence in which "the ridiculous fetters of so-called human feeling" would finally be broken. Schwarz, *Der konservative Anarchist*, pp. 73, 122-124. For Jünger could not believe in a world in which there was no tension, no struggle between opposed choices, especially for the soldier. The true soldier chooses to go on against strong wishes for rest or flight; he cannot always believe and never doubt, nor can he suppress in himself the peaks and valleys of emotional experience. The heroic individual struggles with himself first of all, and that struggle is a moral one.

By contrast, the Nazi conception, as expressed by Hitler particularly was a mechanical one. Hitler also knew about the tensions generated by combat, but this was precisely what he hoped to eliminate through breeding. See, for example, *Mein Kampf*, trans. Ralph Manheim (London, 1969), pp. 151–52. Hitler wanted soldiers who would not experience or express the ambivalent, selfish, and even cowardly feelings often encountered during the Great War. Aryans, he insisted, are incapable of such feelings, and anyone who expressed them suffered from corrupted blood. There was no morality involved in either a bourgeois or an aristocratic sense. Thus, Hitler often ordered troops to hold positions, against the judgment of commanders on the spot, because Aryan will should force the others to break first, and if they were truly overwhelmed then they should fight to the death, as instinct dictated. The failure to respond instinctively was racial by definition, and such men could perish in any case.

Jünger wrote about war as an extreme situation leading to heightened experience, as the individual is absorbed into a collective. In war, man's strength, courage, and capacity to endure are pitted against feelings of fear, fatigue, loneliness, and revulsion. War discloses the inevitable limitations of strength and courage, and the soldier must have the ability to live with contradictory feelings. It was the presumed absence of such feelings, the denial of their validity, that took war out of the realm of moral choice and permitted Hitler to wage a racial war against unarmed women and children. Jünger was appalled by the uses to which the German military capacity was subjected. There was no soldierly virtue in humiliating and destroying the helpless, and Jünger said that he was disgusted by the uniforms and weapons he once loved, and he lamented the death of chivalry and the domination of war by technicians. In Jünger's view, war was in the realm of moral endeavor, war presented men with a dilemma of moral, not mechanical dimensions and consequences. This is why he wrote that Nazism was the "ametaphysical solution," the merely technical execution of his vision of "total mobilization." Schwarz, *Der konservative Anarchist*, pp. 122, 129, 164–166, 180, 191–193; Alastair Hamilton, *The Appeal of Fascism: A Study of Intellectuals and Fascism, 1919–1945* (London, 1971), pp. 167–168; Wulf, *Literatur und Dichtung*, pp. 10–11. It is not hard to distinguish an instrumental or opportunistic kind of repudiation from Jünger's moral repudiation. See, for example, Christopher R. Browning, "Martin Luther and the Ribbentrop Foreign Office," *Journal of Contemporary History,* vol. 12 (1977): 333–339.

32. On Benn's thorough disillusionment with the Third Reich, see also *Ausgewählte Briefe,* letters to Frank Maraun of May 11, 1936, and April 12, 1936, pp. 71, 68; and "Art and the Third Reich," in Ashton, ed., *Primal Vision,* pp. 96–99.

33. See, for example, Joseph Bendersky, "The Expendable *Kronjurist*: Carl Schmitt and National Socialism," *Journal of Contemporary History,* vol. 14, (1979): 309–328.

34. Wilhelm Stapel, *Der christliche Staatsmann* (Hamburg, 1932), pp. 265–66; and *Die Kirche Christi und der Staat Hitlers* (Hamburg, 1933), pp. 13–15.

35. The paramount German goal, according to Stapel, was hegemony in Europe, which should be established by force if others refused to grant

its legitimacy. There was no equality among individuals and certainly none among states. Germans are superior to other people and if that is unjust, then the cause lies with God. Stapel argued that if such power is immoral then God is immoral, but he really did not have to push that logic too far. The evidence for the necessity of his position was immediately manifest in purely human activity. "Look at a marching troop of German youths and realize what God has made them for! They are warriors by nature and their calling is to rule." Wilhelm Stapel, *Der christliche Staatsmann*, pp. 228, 246–250, 252–256, 272.

36. Stapel argued, for example, that right exists in relation to the state, not the individual, and the definition of right was not a matter of agreement but of the leader's will. "The people do not know what they want; they have only instinct. But the leader knows what the people want, and that is what makes him a leader." *Die Kirche Christi und der Staat Hitlers*, pp. 14–16, 26. Stapel had no difficulty accepting the Nazi regime because he viewed the movement in rather traditional conservative, nationalist, and Prussian terms. See his *Preussen muss sein! Eine Rede für Preussen* (Hamburg, 1932), pp. 30–31, 38–39, 43.

37. Stapel's anti-Semitism did not conform by any means to the standards of Nazi racism, being rather cultural and religious. Stapel argued for the need to separate the two peoples but he rejected economic anti-Semitism, for example, seeking "only" to remove Jews from involvement in German political and cultural life. On Stapel's rather tame anti-Semitism, see "Versuch einer praktischen Lösung der Judenfrage," in Albrecht Erich Günther, *Was wir vom Nationalsozialismus erwarten! Zwanzig Antworten* (Heilbronn, 1932), pp. 186–191, especially p. 189. Stapel saw the struggle in Europe as one between France and Germany, the Jews being the agents of the French revolution on German soil (*Der christliche Staatsmann*, pp. 256–257, 259). But the Nazis had no patience with this pale image of "a flood of foreign blood." In May 1937, Stapel was bitterly attacked for his historical version of "the Jewish question," his failure to stress the racial basis of that question, leading to a less than thorough anti-Semitism. Stapel was censured for his insensitivity to "the realities of race and blood," for his notion that Jews might become members of the German community, which showed that Stapel was totally lacking in political instinct. Heinrich Kessler, *Wilhelm Stapel als politischer Publizist* (Nürnberg, 1967), pp. 209–210, 206–208.

38. Kessler, *Wilhelm Stapel*, pp. 219–224.

39. Robert P. Eriksen, "Theologian in the Third Reich," *Journal of Contemporary History*, vol. 12 (1977): 595–622.

40. Binding's statement is in Wulf, *Literatur und Dichtung*, p. 91. Although Binding felt that he had no right to withhold his cooperation from the regime, he nevertheless complained that his name had been used without permission on a list of 88 writers who had signed an oath of loyalty to

Hitler. *Ibid.*, p. 96. On Binding's disillusionment with the regime, see *ibid.*, p. 89; see also the review of Binding's published correspondence in *Die Zeit*, August 15, 1957, p. 15; and *ibid.*, January 8, 1965, p. 9. Bronder, *Bevor Hitler kam*, pp. 85–86. Binding thought, among other things, that *Mein Kampf* should never have been written,

41. Eduard Spranger, "Mein Konflikt mit der Hitler-Regierung, 1933," in Léon Poliakov and Josef Wulf, *Das Dritte Reich und seine Denker* (Berlin, 1959), pp. 89–94; Spranger was a member of the Wednesday Society (*Mittwochsgesellschaft*) along with Ferdinand Sauerbruch, Ludwig Beck (mentioned in n. 42 immediately following) and others in the "opposition." See Spranger's *Berliner Geist* (Tübingen, 1966), pp. 117–127.

42. Hans Carossa, *Ungleiche Welten* (Wiesbaden, 1951), pp. 31, 45, 220, 234–236, 240; on Rudolf Pechel, see Kurt Zeisel, *Das verlorene Gewissen* (Munich, 1962), pp. 19–20, 24–28; on Hauptmann, see Erich Ebermayer, *Denn heute gehört uns Deutschland . . . Persönliches und politisches Tagebuch* (Hamburg, 1959), p. 264; and Wulf, *Literatur und Dichtung*, pp. 132–33. On Ludwig Beck, see Nicholas Reynolds, *Treason was no Crime* (London, 1976). Ina Seidel wrote, for example, "I was not merely smitten with blindness, I was simply a sheep and not even a black sheep, but one of millions." In K. Held (nom de plume), "Prisoners of the Past: New German Writers," *The Wiener Library Bulletin*, vol. 1, nos. 5, 6 (September–November 1951): 25. On Wilhelm Niemöller, see George Mosse, *The Crisis of German Ideology* (New York, 1964), p. 308. See also the chapter on a number of writers in Bronder, *Bevor Hitler kam*, pp. 83–98.

43. C. G. Jung, "The State of Psychotherapy Today," *The Collected Works of C. G. Jung*, trans. R. F. C. Hull, 17 vols. (London, 1964), 10: 165–67.

44. *Ibid.*, pp. 158, 165–67. Jung shortly recovered, of course, having remembered that Wotan was a fundamental attribute of the German mind, which portended great danger. By 1936, references to the "Aryan race" were vulgar, Germany was a land of spiritual disasters, and the "we" referred to Jung's fellow Swiss citizens. C. G. Jung, "Wotan" and "After the Catastrophe," *ibid.*, 179–193, 200.

45. The Jung matter will not soon be laid to rest. Mortimer Ostow, a psychoanalyst, recently published an excerpt from a previously unpublished and unknown letter of Jung's of February 9, 1934: "As is known, one cannot do anything against stupidity, but in this instance the Aryan people can point out that with Freud and Adler, specific Jewish points of view are publicly preached and, as can be likewise proven, points of view that have an essentially corrosive character. If the proclamation of this Jewish gospel is agreeable to the [Nazi] government, then so be it. [But] there is also the possibility that it will not be agreeable to the government." Mortimer Ostow, "Letter to the Editor," *International Review of Psychoanalysis*, vol. 4, no. 3 (1977): 377.

46. Aniela Jaffe, *From the Life and Work of C. G. Jung*, trans. R. F. C. Hull (London, 1972), pp. 97–98.

47. Alec Nisbett, *Konrad Lorenz* (New York, 1976), pp. 81–83, 134. "The 1940 paper tried to tell the Nazis that domestication was much more dangerous than any alleged mixture of races. I still believe that domestication threatens humanity; it is a very great danger. And if I can atone for the (retrospectively) incredible stupidity of having tried to tell the Nazis about this, it is by telling the same obvious truth to quite another sort of society which likes it even less." *Ibid.*, p. 90. Also see Lorenz's statement about when he discovered what the Nazis were doing, *Ibid.*, p. 94.

48. For example, one of the commonplace sentiments of the time was that war is the father of all things, meaning that it is not only necessary and desirable but also enhancing. But this sentiment had a cognitive content (in terms of personal, class, or national interest), a moral content (for those who saw in war an elevating experience of manly sacrifice and comradely discipline), and a wishful content (which justified all kinds of sexual and aggressive as well as narcissistic strivings). Now this kind of statement had a moral significance for a man like Jünger that it never had for Hitler, at least not as the leader of the Nazi regime. Friedrich Hielscher, a friend of Jünger's and an aggressive German nationalist revolutionary himself, visited the Lodz ghetto in 1941, spoke to Jews who were about to die, and wrote, "I felt ashamed of my German name, more deeply ashamed than I ever thought I could be." *The Wiener Library Bulletin*, vol. 13, no. 1, 2 (1959): 16. See also Kurt Sontheimer, *Antidemokratisches Denken in der Weimarer Republik* (Munich, 1962), pp. 289–291.

49. Grandiose or narcissistic entitlement is a concept that I will refer to again in chapter 3, in reference to Hitler. The fact that this boundless seeking is not restrained by moral scruples does not mean that it is not motivated by ideal images—on the contrary, it is this that distinguishes Hitler's seeking from mere nihilism. On entitlement, see Arnold Rothstein, "The Ego Attitude of Entitlement," *International Review of Psychoanalysis*, vol. 4, no. 4 (1977): 409–417.

50. The problem of masking intentions and meaning by the uses of language, especially in this instance in terms of being and doing and terms of abstract or concrete levels will be discussed at length in Chapters 3 and 4. See Edith Jacobson, *The Self and the Object World* (New York, 1964), p. 146.

51. Obviously there were many people who felt unable to effect change or control a desperate situation through political activity as it existed, and they were able to rationalize their allegiance to Hitler and the movement according to acceptable standards, assuming that he was in some sense a traditional conservative, or that conservatives would finally dominate the government, restoring order, protecting them from communism, and removing them from the need to be involved in sordid political concerns.

52. Walther Hofer, *Der Nationalsozialismus: Dokumente 1933–1945* (Frankfurt a/M., 1957), pp. 180–181; Robert J. O'Neill, *The German Army and the Nazi Party, 1933–1939* (London, 1966), pp. 125–127, 39–41.

53. Koppel Pinson, "Freedom without a Conscience," *The Wiener Library Bulletin*, vol. 12, no. 3, 4 (1956): 23, 30.

54. Werner Thomas Angress, "The Response of the Educated Jewish Elite in Germany to National Socialism, 1933–1939," in press. I am grateful to Professor Angress for pointing this out to me. See also Gottfried Reinhold Treviranus, *Das Ende von Weimar* (Düsseldorf, 1968), p. 324.

55. Leo Szilard, "Reminiscences," in Donald Fleming and Bernard Bailyn, eds., *The Intellectual Migration: Europe and America, 1930–1960* (Cambridge, Mass., 1969), pp. 95–96. See also Carl Zuckmayer, *A Part of Myself*, trans. Richard and Clara Winston (New York, 1970), p. 321. "We ourselves went on hoping for the best up to the last moment, and even then we did not want to believe the trend was irreversible. For us, that last moment was the Berlin Press Ball on January 29, 1933."

56. Joseph Haberer, *Politics and the Community of Science* (New York, 1969), pp. 113, 132. Beyerchen, *Scientists under Hitler*, pp. 63, 227, n. 44. Typically, many said that they went along to moderate the influence of the Nazis in science, music, etc. *Ibid.*, pp. 176–177. Wulf, *Literatur und Dichtung*, p. 96. Wulf referred to the 88 writers and poets who signed an oath of loyalty to Hitler, suggesting that the list was not very convincing because some had signed merely to protect their publishers.

57. "What does one do, however, when one cannot untie a knot? One cuts it in two and masters it. . . . In the face of this imperishable accomplishment the fact that it at first believed that it had to destroy many other real values, perhaps unjustly, carries little weight. It is well known that in every great housecleaning some pieces of porcelain are bound to be broken." Adolf Spemann, quoted in George Mosse, *Nazi Culture* (New York, 1966), p. 161. See also, for example, Eva Lips, *What Hitler Did to Us* (London, 1938), pp. 21, 28; Judith Kerr, *When Hitler Stole Pink Rabbit* (London, 1971), pp. 24–25; Hermann Ullstein, *The Fall of the House of Ullstein* (London, n.d.), pp. 14, 197, 223. The Wiener Library has many personal reports and eyewitness accounts that are of the same order, a misreading of events based on common sense observations in terms of interest, morality, or wishfulness. See the volumes of documents, *History up to 1933* and *The Seizure of Power to the Outbreak of the War* (1933–1939).

58. For example, Ebermayer, *Denn heute gehört uns Deutschland*, pp. 11–13, 78–79, 81, 84, 89, 115, 119, 123–125, 155, 175, 200, 203, 210–211, 225–226.

59. *Ibid.*, p. 49; Helmut Kuhn, ed., *Die Deutsche Universität im Dritten Reich* (Munich, 1966), p. 51; Theodor Heuss wrote, upon returning from a meeting with Jewish students in February 1933, that it was as if he had been talking to frustrated Nazis. Quoted in H. W. Koch, *The Hitler Youth: Origins and Development, 1922–1945* (London, 1975), p. 118. Ac-

cording to Carl Zuckmayer, "even many Jews considered the savage antisemitic rantings of the Nazis merely a propaganda device, a line the Nazis would drop as soon as they won governmental power." *A Part of Myself*, p. 320.

60. Gerhard Fricke, in Rolf Seeliger, ed., *Braune Universität: Deutsche Hochschullehrer gestern und heute*, 6 vols. (Munich, 1966), 3, pp. 49–51. Fricke was professor of literature at Cologne University. He participated in the book burning of May 10, 1933, and did his share in celebrating the passing of the Western diseases (historicism and relativism, in this case), and the restoration of the eternal, unconditional reality of the people including, of course, its soil, blood, community, honor, and chosen destiny. Fricke wrote that what is revealed here "is not error or failure but guilt, however unwanted or unwitting. It is open neither to contradiction nor to atonement, and any attempt to minimize our own involvement is doomed to failure." *Ibid*. In 1933, Fricke had urged people to plunge into the movement, serve it, and fight for it, as the only way to share in the mighty current of power. Wendula Dahle, *Der Einsatz einer Wissenschaft* (Bonn, 1969), p. 186. Fricke's efforts through the Nazi period can be traced briefly in this useful work, pp. 147, 164, 176, 186, 212, 220, 233, 246.

61. Charlotte Zernik, *Im Sturm der Zeit: Ein persönliches Dokument* (Düsseldorf, 1977), p. 45.

62. Victor Klemperer, *LTI: Notizbuch eines Philologen* (Berlin, 1949), p. 48.

63. Quoted in Beyerchen, *Scientists under Hitler*, p. 18. See also Hans Egon Holthusen, "Porträt eines jungen Mannes, der freiwillig zur SS ging," in Wolfgang Weyrauch, ed., *War ich ein Nazi?* (Munich, 1968), p. 60.

64. Joel König, *Den Netzen entronnen: Die Aufzeichnungen des Joel König* (Göttingen, 1967), p. 43, 46. Rabbi König's ability to interpret events in the direction of his wishes was remarkable; see also *ibid*., pp. 44, 86.

65. Wilhelm Hoegner, *Flucht vor Hitler: Erinnerungen an die Kapitulation der ersten deutschen Republik, 1933* (Munich, 1977), p. 121.

66. The Nazis had polled 13,765,781 votes in July 1932 (37.4% of the total) and 11,737,000 in November 1932 (33.6% of the total). Bracher, *The German Dictatorship*, p. 159. Hans Mommsen, "National Socialism: Continuity and Change," in Walter Z. Laqueur, ed., *Fascism: A Reader's Guide* (London, 1976), pp. 188–189; Thomas Childers, "The Social Bases of the National Socialist Vote," *Journal of Contemporary History*, vol. 11 (1976): 29–30.

67. Werner Forssmann, *Experiments on Myself*, trans. Hilary Davies (New York, 1972), pp. 126, 141. In addition, see Melitta Maschmann, *Account Rendered: A Dossier on My Former Self* (New York, 1965), pp. 10, 26, 28, 41, 55. See also Joachim Günther, "Rückblick und Rechenschaft," Hans Egon Holthusen, "Porträt," and Hans Winfried Sabais, "Menschenmaterial," in Weyrauch, ed., *War ich ein Nazi?*, pp. 39, 56, 70, 135, 139–145,

respectively. Gudrun Tempel, *Speaking Frankly about the Germans* (London, 1963), pp. 12–13; Arnold Freiherr von Vietinghoff-Riesch, *Letzter Herr auf Neschwitz* (Limburg, 1958), p. 204. Hans Winfried Sabais noted that the youth wanted a communal life and to be challenged; Weimar had thrown this energy away, dissipating it on union activity, allowing it to collapse in apathy. The brown shirts, by contrast, allowed people to feel important.

68. There are many interpretations of Hitler's behavior. Maurits Katan, "A Psychoanalytic Approach to the Diagnosis of Paranoia," *The Psychoanalytic Study of the Child*, vol. 24 (1969):329, refers to Hitler as a narcissistic personality, as does Erich Fromm, *The Heart of Man* (New York, 1964), pp. 66–76. It should be noted that Fromm does not repudiate his earlier reference to the sadomasochistic nature of Hitler's character, but simply encompasses both positions. See his *The Anatomy of Human Destructiveness* (New York, 1973), p. 406; Robert G. L. Waite, *The Psychopathic God* (New York, 1977), treats Hitler in rather conventional, ontogenetic terms; see also his reference to Gertrud Kurth, in "Adolf Hitler's Anti-Semitism," *The Journal of Interdisciplinary History* vol. 1 (1971):228, n. 68. Rudolf Binion, by contrast, rejects the ontogenetic approach and interprets Hitler's behavior in terms of traumatic experience. *Hitler among the Germans* (New York, 1976). Heinrich Himmler has been discussed by Peter Loewenberg, "The Unsuccessful Adolescence of Heinrich Himmler," *American Historical Review*, vol 76, no. 3 (June 1971):612–654; note especially Loewenberg's reference to Harry Guntrip, *ibid.*, pp. 617 and n. 15, p. 617. On Goering, Hess, Hans Frank, and others, see Gustav Gilbert, *The Psychology of Dictatorship* (New York, 1950), pp. 81–153.

69. Theodore Abel, *Why Hitler Came to Power* (New York, 1938); Peter Merkl, *Political Violence Under the Swastika: 581 Early Nazis* (Princeton, N.J., 1975), pp. 273–282. What the data in these works show is that Hitler all along appealed to many kinds of people who expressed in their statements anger, bitterness, envy, anxiety, depression, and desire for revenge. There is no one personality syndrome that can encompass the Nazi leadership, let alone the public at large. Oswald Spengler had said that German socialism must grow out of German nature—and then he and Moeller van den Bruck, the Strasser brothers, Ernst Niekisch, and Ernst Jünger, Hans Blüher, Hans Zehrer, Ferdinand Fried, Werner Sombart, Othmar Spann, and others besides proliferated versions of that German socialism without ever drawing the obvious conclusion; there was no German nature and insofar as there was a "German mind," it was very fragmented and diverse. Hitler also had a version of German socialism, but it was so narrowly cast that none of the others could ever reconcile themselves to it.

70. Henry V. Dicks, *Licensed Mass Murder* (New York, 1972), pp. 63–70. The authors of *The Authoritarian Personality* (T. W. Adorno et al.

[New York, 1969], pp. 762–777), indicated in 1949 that there were several types involved in Nazi activities, identifying what they called rebel, psychopathic, crank, and manipulative types. Ernst Röhm was considered an example of the first and Heinrich Himmler as possibly an example of the last. The authors always stressed the primary role of the sadomasochistic type and its lower-middle class origins, but they could never account for the real complexity, which they recognized, in these terms.

71. Franz Alexander, for example, wrote of the bitter disillusionment of many liberals when they discovered that "Soviet totalitarianism in all essentials was identical with the Nazi regime." *Our Age of Unreason* (New York, 1951), p. 226. But if Nazi totalitarianism came out of the socialization experiences of the legendary Central European family, then what did Soviet totalitarianism come out of? That question will never be answered in terms of familial socialization.

72. Zuckmayer, *A Part of Myself*, p. 319. Paul Althaus, in Rolf Seeliger, ed., *Braune Universität*, 3, p. 13; see also the statements of Gerhardt Giese, *ibid.*, 4, 119–120. It is important to distinguish moral from cognitive orientations. It may be very appropriate to repudiate moral orientations rooted in conflict and to stand for moral orientations rooted in cooperation. But one should not confuse matters by claiming that such a stand is taken in the name of "value-free science."

73. That is, if systematic rationality is claimed for one's own position, holding that any "irrational" manifestations as defined by the standards of the position are random and idiosyncratic, then there is no basis for anticipating systematic irrationality and no basis for analyzing it when it occurs. It is simply held not to occur no matter what happens. Individuals may be implicated in a random way (hence the "cult of the personality," or the trial of Lieutenant Calley), but not the position as such. The point is that all orientations to reality are based on moral commitments, or superego standards, and no orientation can therefore be viewed simply as "rational"; all orientations must be viewed as "nonrational" as well.

74. In addition to Heidegger, Benn, and Binding, others who saw potential moral worth in Nazism included Hermann Rauschning, *Hitler Speaks*, pp. 150, 213–214; Ernst Hanfstaengl, cited in John Toland, *Adolf Hitler* (New York, 1976), p. 540; and Albert Krebs, as discussed by William Sheridan Allen in his introduction to Krebs' *The Infancy of Nazism: The Memoirs of ex-Gauleiter Albert Krebs* (New York, 1976).

2

The Emotional Impact of Events: When the Common Sense World Fails

The Nazi movement was heterogeneous in social composition, not to the extent the Nazis claimed and could have wished for, considering their role as champions of the community, but more than can be inferred from the constant reference to Nazism as a movement of the middle and lower-middle classes, and of the young. In fact, regardless of the category of analysis (class, age, occupation, region, or religion), no responsible author has ever claimed more than the existence of definite statistical *tendencies*, because there were statistically significant numbers of people from the same background who did not support the movement and from different backgrounds who did. The heterogeneous composition of the Nazi movement has been commented on so many times that no significant elaboration is required here.[1] It is necessary only to point out some interesting features of the problem and to draw certain conclusions from them.

In S. M. Lipset's well-known characterization, the ideal–typical Nazi voter in 1932 was a middle-class self-employed Protestant who lived either on a farm or in a small community, and who had previously voted for a centrist or regional political party strongly opposed to the power and influence of big business and big labor. However, Lipset also showed that in 1930 40% of Social Democratic Party (SDP) voters were not manual workers and that the party was backed in that year by 25% of the self-employed in artisan shops and retail business.[2] Lipset did not try to account for the discrepancy nor did he account for any fluctuations over time.*

By the same token, there was working class support for the Nazi Party. This support had greater significance for the shape of the party itself than it had for the relationship of the party to the working class at large. But whatever the qualifications, no serious observer would deny that there was some measurable degree of working class support for the Nazis.[3] And although the political resistance of Catholic voters to Nazi appeals is a well-established fact, especially through the crisis years 1930–1933, there was a significant degree of support for the Nazis among Catholics, as the electoral returns indicate.[4]

In a similar vein, Peter Loewenberg has stressed the enormous appeal of the Nazi movement to the young, explaining this appeal, not as Lipset had done, on the basis of class-rooted socialization experiences, but on the basis of traumatic experiences of a particular wartime generation, the response to which was recovered and repeated in the traumatic situation of the 1930s.

> According to the Reich's census of 1933, those 18 to 30 constituted 31.1 percent of the German population. The proportion of National Socialist party members of this age group rose from 37.6 percent in 1931, to 42.2 percent a year later. . . . By contrast, the Social Democratic Party, second in size and the strongest democratic force in German politics, had only 19.3 percent of its members in the 18 to 30 age group in 1931.

Once again, however, not only did nearly 20% of the same age cohort belong to the opposition party, but the majority of the Nazi party membership obviously did not belong to that particular cohort and cannot be explained in the same terms. Indeed,

---

* The Social Democratic Party was also backed in 1930 by 25% of the white-collar workers and by 33% of the lower civil servants.

28% of the Nazi party membership was born between the years 1873 and 1892, and a measurable proportion (3.6%) was born even before then.[5]

The often-remarked reluctance of salaried, middle-class, white-collar employees to identify themselves and ally with blue-collar workers presents an interesting example of the problem of heterogeneity. Many private (and public) salaried employees joined unions during the Weimar period, though the majority did not. Of the three major unions representing these white-collar employees, one was affiliated with the unofficial labor organization of the Social Democratic Party. By 1928, the other two unions that sought a status distinct from and superior to blue-collar workers had a combined membership twice that of the union affiliated with the SPD. Nevertheless, in 1931, some 25% of white-collar workers were represented by this union, and at the height of the crisis, in 1933, a significant number of employed white-collar workers eligible for membership in the left-wing union still belonged to it.[6]

Obviously, there were a number of left-wing, white-collar, salaried employees compelled by the contemporary situation to give up their socialist affiliation. But they did not all do so, and when they did it was not necessarily because they were predisposed or susceptible to authoritarian solutions, as the following statements indicate:

> I am bitterly disappointed by the attitude of our [Social Democratic] leaders who show so little interest in class struggle at this point. . . . [I] have come to the conclusion that there can be no further progress under their direction. Under these circumstances I feel unable to follow the party line any longer. . . .

> As a civil servant I have to make a choice. On the one hand, I see now the tendency is growing on the part of my employer, the Reich, not to tolerate those employees belonging to anti-government associations. On the other hand, there is my loyalty to the [Social Democratic] Party. Unfortunately, I see no other solution but to resign. The existence of my family is at stake. If the fate of unemployment, which in my experience can be *very, very* hard, is unavoidable I need not reproach myself for not having done everything in the interests of my wife and child.[7]

There are many other such statements indicating an inability to tolerate the increasing radicalization of politics, fears for the future, for the welfare of families, or recognition that the new regime would not tolerate divided loyalties as reflected in SPD membership. To be sure, many white-collar, salaried employees were of proletarian background and maintained proletarian loyalties, which would account for such statements. But 39% of white-collar workers of proletarian background belonged to the right-wing unions.[8]

A threefold problem is raised by the dilemma of heterogeneity. First, young people or people of lower-middle and middle-class backgrounds cannot be treated systematically while other elements of the population, who were also involved, are not. If the tendencies suggested by theory and identified in fact can be shown not to have been accidental, the opposed tendencies can also be shown not to have been accidental. Second, treating the other elements of the population systematically—either those of similar background who did not support the movement or those of different background who did—then requires the multiplication of theoretical entities, always a sign of inadequate theory. Moreover, explanations of the fluctuations in support before the seizure of power and of the regime's ability to mobilize a heterogeneous population afterwards are also required. The Nazi terror was obviously meant to stifle any expression of opposition, but the Nazis could count on the loyalty and even enthusiasm of large numbers of people who were not supporting them just out of fear.* Finally, if statistically significant numbers of people defined objectively in terms of class and age did not behave in expected

---

* The fluctuation in support for the Nazis in November and December 1932 and January 1933 suggests that the population was affected by events and personalities in ways that cannot be accounted for by current theory and in numbers that cannot be anticipated by it. Sociologists have always been aware of this range of problems, of course. They have always acted as if further accumulation of data or further refinement of theory would allow the problems to be resolved. But the numbers will not change and there is no way that traditional sociology can account for the data. In this respect, social science is more interesting for what it cannot explain than for what it can.

ways, despite the fact that by definition they shared determinative experiences, then it also becomes necessary to explain contradictory conclusions on the basis of the given premises. This is beyond the scope of conventional theory.

In short, there were definite, consistent, and expectable tendencies as suggested by theory, but the correspondence is far from perfect, and it is necessary to explain the heterogeneity of response in order to account for the data in the most comprehensive way. The difficulties are obvious, but there is at least one immediate advantage accruing from the attempt: It frees us from certain preconceptions, especially personal or characterological continuity, which follow from notions of class, age, or religious needs and loyalties, or of predispositions to behavior. If statistically significant numbers of people responded in ways not anticipated by theory, then the concept of personal continuity is inapplicable.

It is equally important to note that heterogeneity of response occurs in *all* social movements to varying and inconstant degrees. This fact confirms the uniqueness of historical events, but it also casts doubt on the expectations of social science with respect to precise explanations of both past and future events. Historians have always thought that social scientists, regardless of theoretical orientation, could never achieve the level of precision and consistency ordinarily associated with scientific conclusions, and they have never had any trouble identifying the negative instance with respect to past events. But the invariable recurrence of heterogeneity of response in social movements indicates that there can be no precise predictive social science either.

Karl Popper has explained that

> no deterministically operating computing machine can calculate *its own future* in full detail, and hence the career of a deterministic system will *not* be completely predictable by a human or a computer if they themselves are constituents of that system.[9]

Popper thought this was so because any individual's location in society will inhibit insight into matters that affect his own position.

However, Freud's discussion of affect is even more to the

point. According to Freud, the realistic and moral assessment of danger situations, as well as the wishfulness evoked by them, vary from person to person. No single event triggers anxiety, for example, and no single form is characteristic of it. Freud observed that there are prototypical forms of anxiety derived from different stages of development (e.g., fears of separation, of the loss of love or of the loved object, of castration, and of superego). These different forms coexist in development, and a current experience that resembles an earlier one can force action at any time and at any one of these developmental levels. The form of anxiety is not created anew for every situation, but is reproduced as an affective state "in accordance with a memory picture already present." [10]

Psychoanalysts have since observed that the same holds true as a developmental principle for other affective states, including especially depression, which encompasses feelings of helplessness and hopelessness. Depressive experiences can recur too any time in life: helplessness whenever someone on whom one has become dependent for support or for specific forms of assistance is perceived as lost; and hopelessness when one seems unable to achieve a goal that is important for a sense of self-respect and self-esteem.[11]

Of course, all people experience developmentally a variety of threatened or actual situations of loss, separation, deprivation, or neglect. People lose those close to them whom they have loved, and they also lose more abstract symbols, culturally important individuals, and perhaps even narcissistically valued character traits that support socially useful skills and moralities. Thus, when social conditions assuring sublimated forms of gratification or protection from adverse circumstances are not met (as, for example, when employment is unavailable or when the authorities perceived as responsible appear threatened, indifferent, or inadequate), some affective experience such as anxiety or depression is bound to result.

However, the perception of such dangerous situations and the assessment of one's adequacy in the face of them vary from one person to the next, according to the unique vicissitudes of life. People will therefore respond differently to the same situation (interpreting events variously as unjust, humiliating, depriving and being angry, depressed, anxious, and so forth). How people feel about situations and how they assess them are not determined exclusively by objective location in society. This is why social movements are heterogeneous in composition, why responses cut across class, occupational, regional, religious, and other lines, and why the numbers change as conditions appear to change.

The predictive capacity of social science is problematic because no one can know beforehand what areas of society will be disrupted by events that are, more likely than not, unanticipated from the standpoint of the conscious expectations of people. No one can tell how people will respond to such events, whether they will feel injured or personally disrupted or elevated and enhanced, as workers, parents, citizens, comrades, or students. No one can tell with any degree of precision which way people will go, or if they will go at all. Doris Lessing's observation that the future is what happens is possibly overstated. It is possible that the social sciences will be able to predict *tendencies,* though I am not confident that they can routinely do so in situations as complex as the breakdown of Weimar Germany.

The Importance of the Concept of Affect in Theory and History

Freud's discussion of affect rendered problematic the traditional conception of sociology. However, by elaborating a theory of affect, Freud provided the basis for a psychoanalytic sociology, particularly important for understanding the problems of heterogeneity and continuity, as discussed in the preceding section and in Chapter 1. Freud assigned a regulative role to affective sensation, which he conceived of as "signaling" impending dangers, that is, as part of a regulative system capable of initiating self-protective actions based on memories of earlier threatening experiences. This broadened psychoanalytic concerns to include ontological insecurities compelled by social relationships and disruptions of social life.[12]

Freud had suggested that a variety of urgent or compelling

thoughts might arise that are not motivated by libidinal or aggressive strivings but by disrupted social relationships. Moreover, adaptive and defensive behavior mobilized as a result of sexual and aggressive impulses, or as a result of the effects of disrupted social relationships, have in common affective experience.[13] That is, when individuals are unable to encompass the significance of events or assimilate all the relevant information, or when events are too threatening to be gathered into and mastered by a personal sphere of competence, self-protective modes of behavior, "signaled" by affective sensations, will follow.[14] Affect is motivational in this sense and becomes a crucial variable in the analysis of social action.

Affects may be understood therefore to constitute the intrapsychic signals for "safe" and "dangerous" behavior, serving to bind interaction as people organize their behavior in terms of accurate or distorted interpretations of emotional expression in others. Most important, the emotional significance of thought and action, certainly an important aspect of the self-evaluative capacity to judge the adequacy of thought and action, develops in a social context. Affect does not arise from within as if the mind is isolated from its environment: Affect, emotional expression, must be learned as a communicative system, just as language is learned, from interpersonal and other environmental interactions.[15] Even those who "feel nothing" as a result of some pathological condition learn and abide by the emotional significance of action for others, lest they be judged "sick" or "deviant" or be overwhelmed by feelings after all, having been rebuked for engaging in offensive behavior.

The emotional significance of thought and action is an effective principle regardless of one's orientation to reality. It may seem paradoxical but it is nonetheless true, as I suggested earlier, that the cognitive orientation we take for granted, or rationality as a technique of environmental control, are of a normative order to which people are emotionally committed. Rational behavior is based on nonrational (unconscious) processes; people may be morally committed to technical development predicated upon the primacy of cognitive functions, but they are not disposed to examine the basis of that commitment. Situations of this sort are typically explained in terms of the concepts "internalization" or "socialization." It is certainly the case that patterns of behavior are transmitted in highly emotional (sensual) contexts that inhibit insight into the source and significance of behavior. But it is also true that people can bring aspects of their world under control; they can manage realistic assessments of one kind of problem or another. Why then can people not bring the world entirely under conscious control, why can they not manage realistic assessments of the full range of experience? Or put another way, why does the internalization process remain effective, inhibiting insight into very important aspects of everyday life—even when society mandates the primacy of cognitive processes? The reason that internalizations remain effective is that people cannot assimilate in consciousness the conflicts generated by the wishful contents involved in their efforts to attain cognitive and other forms of mastery, or by their "developmental crises," nor can they manage either the anxiety or the sadness that would follow from the conscious, integrated acceptance of the random and accidental nature of life. Thus, the cognitive orientation to experience characteristic of our world—no matter how useful it has proved to be with respect to technical mastery—remains at bottom a morality with all that this implies in a psychoanalytic sense, that is, a superego standard that people employ to define their everyday reality and that, for the reasons given, they are compelled to defend. We know from German experience in the 1930s and from our own more recent experience that affective orientations, except in segregated religious, artistic, and sexual forms, are still threatening to people, although as a moral orientation to reality rationality is no better or worse than affectivity, only different.

The emotional significance of action holds true even for the most highly developed rational endeavors, such as science and mathematics. The standards for procedure, method, and proof in these disciplines are also in a normative order to which one is emotionally bound. Society rewards and scientific ideology confirms such apparently objective "affectless" behavior. But such behavior still involves ideal representations of self and others; action consistent with these representations is pleasurable, and violation leads to some felt experience of "pain." What any individual does in this context too must "feel right." [16]

Affects, then, acquire a control or "signal" function, serving to warn of imminent or prospective danger (anxiety) or enabling one to recognize and tolerate sadness and disappointment, a result of realistic limitations and the possibility of failure (depression).[17] A theory of affect becomes indispensable to systematic thinking about groups because the concept is tied to the sense of continuity, to self-representations, to feelings of well-being and safety with respect to socially adequate, shared behavior.

The classic psychoanalytic view of motivation as "arising from basic, unconscious, tension-producing drives" is of no use in systematic sociology.[18] By contrast, taking organized stability and the potential for disrupted stability as a baseline, a theory of motivation based on affective responses associated with the sense of continuity is sociologically relevant because such continuity is defined by the ability to produce socially appropriate behavior based on shared standards and expectations that, like patriotism, heroic strivings, or martial ardor, are not restricted by factors of class, age, religion, and the like.[19]

The kind of mass behavior and affective expression that occurred in Weimar and Nazi Germany, which are inexplicable either in terms of drive or conscious interest, can be accounted for in these terms. That is, when social conflict interrupts the possibility of producing familiar, socially appropriate actions, the sense of "rightness" and hence the sense of adequacy is undermined. The self-critical individual strives to evaluate behavior (the expression of affect, of traits, defenses, and modes of adaptation, including reality testing, cognition, and synthesis) in terms that he has learned (i.e., correct–incorrect, reasonable–unreasonable, true–false, appropriate–inappropriate). However, all this is related to *utilitarian, worldly, everyday ambitions, expectations and goals;* and the inability or feared inability to act, confusion with respect to norms or acceptable modes of behavior, the inability to see oneself in the future, and factors of this order lead to the expression of affect characteristic of failed, threatened, or otherwise unstable ego processes: anxiety, anger, sadness, bewilderment, helplessness, disappointment, purposelessness, and so forth.

Thus, when disruptive social conflict occurs, when routine adaptive activities are interrupted or are perceived as dysfunctional from the start, people are emotionally aroused.[20] This concept of affect, based on a mismatch between novel social demands and a stable repertoire of behaviors threatened by them, on an interruption of "plans" or of "an average expectable environment," involves the perception of real events and of one's adequacy in the face of events, and a search for relief, for the personal and social means of bringing novel events under control. For if events are not controlled, affective expressions will become more global, less focused, and threaten finally the hold on reality.*

The Psychosocial Bases of Hitler's Appeal

Hitler and the Nazi movement became crucial for large numbers of people, particularly after 1930, because they addressed themselves to an immediately disrupting, potentially chaotic situation that had threatened or thwarted the practical exercise of skill and the realization of ideals on an unprecedented and unanticipated scale. Hitler identified and exploited the problem: Indeed, he hammered away at the themes of weakness, passivity, humiliation, defenselessness, promising an immediate reversal, active mastery rather than passive submission, reassuring people that they would not be forced to surrender (again) or be locked in a hopeless battle with the representatives of a hostile and depriving reality.

Thus, it was not simply a matter of order or a disparagement of liberalism, or a refusal to accept the vicissitudes of abstract economic and political processes, that accounts for whatever level of support the Nazis could muster. There were other, more tra-

* In technical terms, "An uninterrupted supply of unperceived, movement produced stimulation appears to be essential to insure effective accommodation to changing environmental conditions. The guidance provided by feedback is most of the time neither obvious nor in one's awareness, becoming so only when the appropriate feedback is no longer available. . . . These circumstances set off an anxious battle to obtain adequate feedback, suggesting that primary danger and profound anxiety are experienced not only in the eruption of drive and in drive conflict, but in conditions in which conception and perception do not fit the environment in a manner that makes action possible or effective." [21]

ditional, "respectable" parties that stood as well for these aspects of German political aspiration. But we cannot understand why those parties also failed unless we have some sense of the impact of events after 1929 or 1930, when millions of people were unavoidably confronted daily with evidence of weakness and inferiority, when the fear of being degraded and depreciated was coming true. And if a lot of people did not believe everything the Nazis said about capitalism, Jews, Bolsheviks, November criminals, and the like, the society was nevertheless in a period of chaotic economic and social decline that was real and had to be faced.

People lived with a sense of failure, of weakness; they were afraid that they would not be able to achieve goals, aims, and ideals (e.g., to be competent and competitive, to be active, masculine, admirable), that present difficulties could not be surmounted, no matter what one did. There was a feeling that Germans were doomed to be victims of merciless forces that they could neither fathom nor control, a situation that was too readily defended against by seeking out challenging and dangerous situations. Thus, the uniformed, marching columns of storm troopers (SA), the martial spirit, the street brawling were exhilarating, integrating experiences for many, especially in that they recalled memories of wartime comradeship, the front-line experience, which was constantly invoked. These activities were in any event more tolerable than the abstract political and economic difficulties, which were remote, alien, and unreal, though possessed of an endless capacity to disrupt and humiliate.

The Nazis, of course, were not remiss in exacerbating the plight of Weimar, adding as much as they could to the sense of chaos they had promised to end. The pitched battles in the streets, the provocations and manipulations, were calculated to demonstrate that the Nazis had the force and the ruthlessness to control the situation, if the Republic did not. There were, obviously, many remarkable instances of brutality. But for those who saw Weimar as betraying the active, masculine, military spirit, fostered and fed by the exigencies of Germany's historical position in Central Europe; for those who found intolerable the feeling that they were not as good, brave, and strong as they had been led to believe; for those who were intent on pursuing a litigious search for compensation as repayment for unjust deprivation; and for those who saw external events and realistic injuries as the source of painful, persistent discrepancies, this was a promising—or at least not a threatening—situation. It has often been said that Hitler fooled especially these Germans, who were gullible and politically naive, lacking political experience. But a gullible person, after all, is one who interprets the world in terms of his need for safety and security without adequate regard for what is realistically possible or plausible.

What this meant, taken all together, is that it had become too difficult for masses of people to evaluate the adequacy of their actions with respect to the present and the future. Against this Hitler and the movement were able to present themselves as strong, fearless, intact, undefeated, and uncorrupted, apparently in control of themselves, organized, and violent. The Nazis could therefore provide all kinds of people with a sense of integration and purpose that was otherwise missing—*in fact*. Against the haphazard and unsuccessful course of Weimar politics, the Nazis offered real and immediate support; against the seeming indifference and even betrayal by Weimar politicians, and against the continued snobbish exclusiveness of traditional conservative parties, the Nazis offered a place and a leader; and against the feared experience of further loss, the Nazis provided a sense of protection, comradeship, and the right to belong. For these reasons people became excited by the impression of power and the promise of renewed activity, interpreting the promise in various, even contradictory, ways.

At the same time, however, the movement was bound to be heterogeneous in composition, for while class, confessional, regional, and other commitments provided a continuing sense of stability and a consistent orientation to action, particularly for working class and Catholic parties, the force of the specific situation was such that *any* member of society could experience a sense of futility, desperation, hopelessness, or rage, considering the Nazi movement consistent with needs, aspirations, and ideals as they were conceived of at that point. The very diversity of the narcissistically valued aspirations that appeared threatened to all sorts of people otherwise distinct in their social location guaranteed the heterogeneous composition of the movement. The situation was bound of course by sociologically and psychologically conditioned limits. But it would have been surprising indeed if even strongly held loyalties to class, religion, region, and the like did not buckle to some measurable degree under the strain. In this way, the Nazis did become something of a *Volkspartei,* as their appeal cut across various social lines.

The relationship of disruptive events to social action is particularly clear in Germany, given the results of the Great Depression. We need only remember that the Social Democrats, the party most committed to the Weimar Republic, was the principal winner of the Reichstag elections of May 1928. A Social Democrat became Reich chancellor for the first time in 8 years, and the party at that point increased its vote by more than one million. Moreover, the appeal of the right-wing parties declined accordingly, in confirmation of the relative stability achieved by the Republic in the preceding period, 1924–1928. In 1928, the share of the total vote among right-wing parties was lower than it had ever been before. The German Nationalist Party did badly, for example, and so did the Nazis, winning but 2.6% of the vote (810,000 votes) and 12 seats in the Reichstag, as compared to the 152 seats held by the Social Democratic Party.

To be sure, the Nazis had by this time developed a strong infrastructure, and their party was well organized, disciplined, and highly motivated; and they could always take advantage of fears prevalent in Germany or stimulate interest in their own program, as they did with their opposition to the Young Plan, which required Germany to continue reparations payments, confirming the hated Versailles Treaty and raising the spectre of ongoing passive fulfillment of the wishes of hostile powers, as it made evident the lack of control Germans felt they had over their national destiny. But through 1929, the Nazis were distinctly a minor party. The upward drift in their favor in different provincial areas is explained by the tight organization, the manipulation of local problems or of local perspectives on national problems, as with the agricultural depression or responses to the Young Plan. In this connection, their alliance with the German Nationalist Party (DNVP), a right-wing bourgeois party, gave them for the first time a national platform through the publishing reach of Alfred Hugenberg, the leader of the DNVP. In no case did the Nazi appeal in 1929 (or before) have the same force that it manifested after 1930. The effects of economic dislocation were unanticipated, as were the gains registered by the Nazis in the Reichstag elections of September 1930.[22]

The results of the depression in statistical terms are too well known to require much discussion here. It has been estimated that in 1932 one in every three of the working population had no job, a situation that of course threatened all those who still had jobs. In any event, there were six million unemployed in 1932 and early 1933. The results in terms of despair, anxiety, and anger are harder to calculate but obviously of the greatest significance. The society appeared to be in a state of chaotic decline. There was constant brawling, in the streets, the universities, even in parliamentary assemblies. Political murder had become a part of everyday life. Hundreds of people were killed or seriously injured in pitched battles during the summer of 1932. On July 17, 1932, for example, 7000 members of the SA marched through the working-class quarter of Hamburg-Altona. The Communists opened fire on the marchers, who returned the fire: 17 people were killed and 65 wounded in the incident. The society had not disintegrated to a state of civil war, but it was not far from it.[23]

There is no difficulty in documenting the sense of despair, anxiety, and alarm that pervaded Germany after 1930, as reported in diaries, journals, correspondence, and the like, a direct result of these events, poignantly captured in the phrase "optimism had grown weary." Ernst Robert Curtius described his return to Heidelberg in 1931, after a long stay abroad: "I was overwhelmed by the darkness of the streets and the inner confusion and insecurity which were evident from the first steps that I took." Germany shudders convulsively, Curtius wrote later, and there was only one hope: "Things must get better because they can't get worse."[24] This was confirmed by Harry Kessler, who wrote at the end of 1931: "A melancholy New Year's Eve, the end of one catastrophic year and probably the beginning of an even more catastrophic one . . . but to bed before midnight in the most dismal state of depression." [25] Leopold Schwarzschild noted in his journals that 1931 was characterized by the fact that attempts to overcome the crisis had only made it more acute. "The same process threatens almost certainly to continue in 1932"; and 1932 was in fact a dismal year, "a frenzied rush towards catastrophe"—until suddenly, at the last minute, a change for the better, "and one would have to be totally lacking in nerves and instincts not to sense it, even in imponderables." [26] The economy had stopped its downward spiral for the first time toward the end of 1932, and the Nazis had for the first time lost electoral support, itself a sign of the cathectic instability of the time. As Schwarzschild's hopes went up, Goebbels' hopes went down: "The year 1932 was one long succession of bad luck . . . I am sitting alone at home, brooding over so many things. The past was difficult, and the future looks dark and troubled; all prospects and hopes have vanished." [27] However dishonest Goebbels may have been in his reporting, this statement seems fair enough: Despair and dismay were rampant throughout the Party in the fall and winter of 1932.[28] Or as one hitherto hopeful conservative expressed it on December 31, 1932:

> This year has robbed us of a great hope—this *year,* not death. Adolf Hitler. Our reviver and great leader for national unity . . . and the man who in the end turns out to be the leader of a party sliding more and more into a dubious future. I still cannot come to terms with this bitter disappointment.[29]

Depression was the most often reported affective experience, and it is no coincidence that the protagonist of Erich Kästner's *Fabian, die Geschichte eines Moralisten* (1932) should have declared that he had been forced to give up hope.

> Whoever is optimistic should despair. I am melancholy; not much can happen to me.

But people everywhere were tired, angry, scared, and anxious because the future had become too problematic. Werner Heisenberg recalled that radical groups . . . demonstrated in the streets, battled each other in the backyards . . . and agitated against one another in public gatherings. Almost imperceptibly the unrest and with it the anxiety, spread also into university life. . . . For a time I tried to push the danger away from myself and ignore the incidents on the streets. But in the end, reality is stronger than our wishes.[30]

And as Erna von Pustau told Pearl Buck,

the desire to come back to earth, to know what life is, the impossibility of knowing where to grasp it, how to approach it, the desperation of old ways lost without seeing any new way on which I could put my foot brought me to a kind of nervous breakdown.[31]

Any individual can continue to struggle on condition of being able constantly to elaborate a fantasy of "some day." Any particular day may be bad, and often is, but as long as one can count on "some day," there is reason to push on. When that is threatened or appears lost, the different, unusual, often alarming responses follow. This was a common experience, of course, but it was particularly hard for the young, which is why they turned to the Nazis. The Nazis seemed able to master the future, which was slipping away. The young in Germany lived "full of concern for the future," which it saw "dark and gloomy, both for [themselves] and the Fatherland." Hence it was that "thousands of young academics who come from poor or ruined families see themselves faced with a precarious future despite talent and hard work, and therefore have no faith in themselves or their world."[32]

This affective expression is crucial because it emerges from and is an explicit response to an immediate, threatening, and objectively damaging situation. People expressed how they felt in a way that does not require, as symbolic or sublimated statements do, a distinct level of analysis and interpretation as the precondition for further analysis and interpretation. The disruption of culturally valued standards and expectations, the threatened or actual inability to realize ambitions or employ valued skills gave rise to unusual and unwelcome intensities of feeling and wishfulness. People scanned the environment for ideas and practices that could answer to their everyday needs and that could also be used to rationalize the unexpected depths of feeling in order to bring the world under control. Worried or frightened by the inability to produce familiar actions, to interpret reality adequately, to be effective as workers or citizens, or otherwise to live up to personal and national ideals, they sought out ideological and organizational ties that could serve to restore in their own minds, with whatever degree of distortion required in any given instance, what was valued, making this experience comprehensible, strengthening thereby their ties to reality. Threats to the sense of adequacy and efficacy, to self-control, personal integrity and the sense of continuity can inspire feelings of ruthless intensity, reflecting the wish to remain active and in control, just as the inability to do so can lead to bewilderment or to the most profound apathy and dejection, or to an overwhelming sense of anxiety. Otto Braun, the Social Democratic prime minister of Prussia, who had been hammered into exhaustion, noted sadly:

> My children are dead, my poor wife has been partially crippled for the past four and a half years and is gravely ill at the moment so that my life for years has been a shuttle between office, parliament and sickroom. This has worn me down terribly and . . . to an alarming degree has drained away my reserves of physical strength.[33]

And the novelist Jochen Klepper, who was married to a Jewish woman and who would, in 1942, commit suicide with her and one of her daughters rather than be shipped off, wrote on March 11, 1933:

> There is a frightful feeling of uneasiness, of pressure, of isolation—a frightful [sense of] weakness, a frightful anxiety affecting existence itself . . . on all sides, escape into private life.[34]

This kind of emotional response to a persistent social crisis, the resolution of which, either with respect to form or timing, was not certain, was repeated any number of times. It was this unstable and exhausting process that was interrupted by Hitler's accession to power. To be sure, this was as much a result of political manipulation as it was of public acclaim, but from either standpoint it was a solution. Hitler was an obviously powerful national leader who seemed capable of restoring public order. And he seemed safe because he was surrounded by conservatives who would make sure that little of the social or racial radicalism of the Nazis would find expression in public policy.

Things did not work out that way because Hitler was determined to pursue his own aims and because people were in no position to seek the gratification of realistic aims through routine solutions. The techniques conservatives tried to use to maintain control, for example, the exploitation of normal channels of information, were ignored in favor of promises of protection and power from an authoritative figure who seemed able now to surmount the feared passivity and to make cherished ideals appear worthy and attainable again. Thus, there followed a great outburst of euphoric jubilation, great expressions of relief, of anticipations of renewal, rebirth, rejuvenation, a great sense of cognitive, moral and wishful enhancement from a public more focused on inner wishes and needs than on realistic aims and routine solutions. Thus, the woman who expressed such bitter disappointment in Hitler and the movement on December 31, 1932, expressed quite different feelings on January 30, 1933:

> And what news did Dr. H. bring with him? His living image, Hitler, is Chancellor! And what a cabinet!!! One we wouldn't have dared dream of last July. Hitler, Hugenberg, Seldte, Papen!!! Each one of them carries a large part of my hopes for Germany. The inspiration of National Socialism, the common sense of the German Nationalist Party, the unpolitical Stahlhelm and Papen, whom we have never forgotten. It is so unbelievably marvellous that I am writing it down at once, before the first sounds of discord. For after the most promising spring, when has Germany ever experienced a summer of fulfillment? Only under Bismarck. . . . A colossal torchlight parade past Hindenburg and Hitler of National Socialists and Stahlhelm who at last, at long last, are marching together again. What a memorable 30th of January.[35]

Or as Hans Bernd Gisevius wrote, "A wild national jubilation erupted, the glorious sensation of a new fraternity overwhelmed all groups." [36]

This national experience was not universally shared, but it was very widespread and it was fostered and manipulated by Hitler, who worked to consolidate power by directing public attention to the recovery of traditional ideals, though he hinted as well of the possibility of release into new realms of experience. Thus, Hans Naumann, professor of literature at Bonn University, participated enthusiastically in the book burning ceremony of May 10, 1933, calling for a German literature "which treats as holy the pious bonds of family, homeland, volk and blood, indeed, all the revered bonds." Naumann called for a literature that was sacred, for poets who would not hide in their studies but would come out, touch the people, and serve as their conscience. He called for a return to purity, chivalry, unity, discipline, and order. Naumann believed in all the traditional pieties and he thought that Hitler, as the leading representative of the National Revolution, did too.[37] Naumann's colleague, Eugen Lüthgen, an authority on philosophy and law, also spoke to the students about the purifying flames, which consumed poison, filth, and trash. Lüthgen expected the whole profound world of German feeling, the tender impulses of the German temperament to be realized as soon as the anti-German currents were blocked and pushed aside.[38] People spoke in those days of the spirit of Prussia and Potsdam, the family, positive Christianity, the soil, the martial spirit, the restoration of dignity to all the traditional roots of German culure. Martin Heidegger invoked the flame at a summer solstice festival, bidding it "to show us the way from which there is no return." [39] The tone was unfailingly grandiose and exalted as people anticipated a breakthrough to "new historic space."

> The violent storm of the national uprising has carried the German people on rushing wings in frontal assault over a thousand years of history to a new stage of development. The Führer calls: a door to great things is open before us—may we not be found unworthy.[40]
>
> Tomorrow has now become today; the mood of the end of the world has changed to one of a new beginning; the final goal be-

comes visible in our time. . . . Deep down in our nation all the forces of a past longing have come alive, and the dream images in which the past indulged have been drawn into the light of day. . . . The new Reich has been created. The Führer, longed for and prophetically predicted, is now here.[41]

A new age has dawned. . . . Colossal upheavals in all sections of German life are leading to a complete transformation of German existence. A united and indivisible Germany has arisen. The work of Bismarck has been crowned. It has all been achieved by a simple man of the people.[42]

With the victory of National Socialism the limitations upon the dynamic spirit of man, which at first frightened us, are transformed into a feeling of restfulness. Freedom . . . is renewed as it rises upward from a close knit consciousness of volk and culture; our world of feeling has found fulfillment.[43]

When one has lived through the political events of the last few weeks and has grasped the deep, powerful meaning of the Führer's passionate words—how he will bring honor and sincerity to international relations, so that lies shall no longer masquerade as truth and truth no longer be dishonored and distorted, how we Germans too shall regain our honor before the world—then the cry breaks out: Here, thank God, are the foundations of a new integrity, here the great, sacred words are beginning to regain weight and currency.[44]

There were literally thousands of such excited expressions, a sign not only that a bitterly disruptive crisis had been resolved, but that loss had been turned to gain. The rapidly unfolding events seemed consistent with German aspirations and well under control. There was no adequate historical or social sense of things because the only thing that counted was the effective interruption of failure, the step from passive acceptance to active mastery. The German public was not concerned at that point with the process; it did not matter how they got from 1914 to 1933. It only mattered that there would be no further losses, that they might even control their own destiny again, and that they had a leader who could achieve all this for them.

People said that Hitler's mission was God-like, that Germans were now prepared to renounce all intellectualism, "not only to serve such a Führer but to love him." [45] Hitler was not less than the Platonic Ideal; on the contrary, he was more because he was real. Hitler was German nature incarnate, the most wonderful personality of all time, a savior (the nameless corporal and the nameless carpenter), a man whose acts could not be judged by the living because they transcended their own time. Hitler was the equal of or superior to Christ, Socrates, Demosthenes, Caesar, Napoleon, Luther, Bismarck. The Reformation had been only half completed and Hitler would now finish the task. Nazism was the dawn of a new era, the old world had collapsed, Germany would show the way to new forms of life, the movement heralded changes of Copernican dimensions.[46]

This exalted language was the result of affect-dominated responses to the environment, which may also be identified by perseveration on idealized virtues, traits, and activities, once threatened and now secure, in confirmation of the narrowed gap between narcissistic aspirations and the capacity to realize them; and by perseveration on the idealized virtues of those who appeared responsible for turning the situation around, and particularly Hitler, who expressed concern for the well-being of the community in a familial way. These ideas were linked by a logic determined by affective states rather than by an ordered perception of the course of events or some insightful understanding of the intentions of the people involved. Information and memories were processed according to their usefulness for maintaining personal stability and the sense of continuity, and while this meant a release from common sense standards for some, most people assimilated the import of events in terms of these standards.[47]

Hence, Germans typically interpreted Nazi demands for duty, discipline, sacrifice, and order as familiar and traditional and therefore as worthy of being attended to. The Nazis seemed able not only to restore but to enhance cherished ideals and conventions. Having once been so diminished and threatened, and now apparently so elevated, people were in no position to assess information or absorb criticism; they did not see any contradictions and they focused loyal and inspired attention on the leader and his colleagues who seemed to have rescued them, as they also focused anger on those whom they identified with their former condition, or at least they condoned or remained indifferent to its expression.

Once the Nazis took power, their orientation to reality was not only segregated from public criticism, it became normative for the society. The Nazis defined "normal" and appropriate behavior, lending a particular structure to the expression of interest, morality, and wishfulness, and they did not lack for doctors, psychologists, ministers, and other authoritative people to confirm the validity of their views. The Nazis had a very great effect over a wide range of activities. They held the keys to status, preferment, prestige, and future chances, and anyone who wanted to move ahead, or even merely to work, had to take that into account.

Of course, not everyone cared to be integrated in the community and the Nazis took no chances: The terror was meant to make opposition appear hopeless to all those who wanted to resist integration. Moreover, there were opportunists, seeking advantages or safety, who could mouth any of the Nazi slogans while believing none of them; and otherwise "normal" people were recruited without great difficulty to serve in the government, industry, and scientific and military establishments, even though they had no great attachment to the movement and might well have disapproved of it. It is indisputable that by January 1935 two-thirds of the party membership had joined after the Nazis came to power, while only some 5% had joined before 1931, when such a choice was especially difficult and even dangerous to make.[48] There were also still some real constraints on the degree to which the Nazis, particularly racial Nazis, could impose their wishes, largely the result of deeply held attachments to conventional morality.

Nevertheless, when opposition, opportunism, passivity, and indifference are accounted for, there was still a remarkable degree of loyalty, enthusiasm, and admiration for the Nazis, cutting across all social lines. The Catholic bishops issued a joint statement ending restrictions on Catholic participation in the Nazi movement at the end of March 1933, and Catholics were only too eager to prove that their faith was compatible with Nazi ideology and that they could contribute to the National Revolution. But the workers too, for the most part, were just as impressed by the economic recovery and just as exhilarated by the stunning foreign policy triumphs of the late 1930s as the rest.

True, the Nazis had all along promised each social interest an advantage that could have been realized only at the expense of all the others, as Otto Braun observed at the time. But it would be difficult to account for the degree of loyalty the Nazis were able to command from 1933 to 1939, and even through the war, just in terms of interest. The Nazis were able to exploit deeply held feelings of devotion to the nation, which were obviously not restricted by class, age, region, or religion. There were contradictions in the society but there was also a high degree of tolerance for living with them, bolstered finally by the needs of wartime. Perhaps the most remarkable example of this tolerance for contradictions was the conviction held by both German conservatives and racial radicals that Nazism represented them primarily, and that one network of standards and expectations would be realized and the other abandoned.

The Problem of Continuity

People everywhere need to maintain a sense of continuity, a sense of self-sameness over time, they need to feel that they are acting on the basis of a "structured" self. Freud's description of how people adjust to neurotic symptoms is useful for understanding how they adjust to the complex movement of everyday life, interpreting and reinterpreting the significance of events in order to maintain a sense of continuity. Screen memories and other "tricks" of memory, the variable capacity for recall or resistance, and so forth, are normally and routinely employed in the service of both tasks. Any individual must be able, through memory, to use time as an ordering principle for the succession of events that constitute his or her history. And as Freud observed as long ago as *Totem and Taboo*:

> There is an intellectual function in us which demands unity, connection and intelligibility from any material, whether of perception or thought, that comes within its grasp; and if, as a result of special circumstances, it is unable to establish a true connection it does not hesitate to fabricate a false one.[49]

"Special circumstances" typically meant to Freud the threatened or actual breakthrough of sexual or aggressive impulses that could finally be controlled only by some kind of symptomatic expression. Should such expression nevertheless become integrated in character (i.e., as a nonconflictual character trait), the sense of continuity would thereby be sustained. Defense mechanisms obviously serve to facilitate this process, even though such mechanisms may be effective at the cost of cognitive and other forms of distortion. But if these mechanisms fail in their purpose, and the symptomatic expression appears alien, the sense of continuity is disrupted and one must either suffer the consequences or be forced finally to seek some form of therapeutic intervention.

The impulsive, threatening, and distorted sexual and aggressive ideas and practices Freud explored are one source of a disrupted sense of continuity. Erik Erikson suggested a second source, noting that different, changing maturational and developmental demands are made on the individual throughout the life cycle. A "developmental crisis"—*holding the social order relatively constant* [50]—may in part be understood as a consequence of the need to redefine identity, or reestablish the sense of continuity, as a result of changing, phase-appropriate maturational demands.

In addition to the psychoanalytic emphasis on "internal" sources of disrupted continuity, there are also various social sources, which psychoanalysts have tended not to explore but which require of people constant defensive and adaptive activity. One is the random and accidental nature of the world, which has its chaotic elements for any individual acting in it. The inevitable proximity of death and vulnerability to traumatic interruptions in daily routines must impinge on psychic processes.[51] Another is the variable and changing forms of social and cultural life, which, though structurally derived and legitimated by the moral order, are nevertheless differently interpreted and integrated in imagination according to any individual's unique experiences.

Finally, the sense of continuity can also be disrupted by failed or failing social standards rendered inapplicable or dysfunctional by violent change, as occurred in Weimar Germany. However, while symptomatic expression or accidental events involve unique instances in which individuals fail to produce expected behavior or are surprised and defeated by isolated events, and while the kind of maturational crises Erickson described involve phase- or age-appropriate responses, the failure of shared standards requires shared solutions that cut across age lines. The sense of continuity, of self-control and efficacy, the ability to act on culturally valued skills and ideals—the disordered subjective world—must be restored by shared activity, organized and legitimated in a social context by forms of leadership and ideology that appear adequate to a solution of the crisis.[52]

Notes

1. The heterogeneity of movement membership and support has most recently been discussed by Barrington Moore, Jr., *Injustice: The Social Basis of Obedience and Revolt* (New York, 1978), pp. 400–411.

2. Seymour Martin Lipset, *Political Man: The Social Basis of Politics* (New York, 1963), p. 148.

3. Moore, Jr., *Injustice,* 406–407; Conan Fischer, "The Occupational Background of the SA's Rank and File Membership during the Depression Years, 1929 to mid-1934," in Peter D. Stachura, ed., *The Shaping of the Nazi State* (London, 1976), pp. 131–159.

4. Walter Dean Burnham, "Political Immunization and Political Confessionalism: The United States and Weimar Germany," *Journal of Interdisciplinary History*, vol. 3, no. 1 (Summer, 1972) :1–30.

5. Peter Loewenberg, "The Psychohistorical Origins of the Nazi Youth Cohort," *American Historical Review*, vol. 76, no. 5 (December, 1971): 1457–1502, especially 1470. In numbers, there were approximately 80,000 people under the age of 25, and 325,000 people under the age of 35 in the Social Democratic Party. Erich Matthias and Rudolf Morsey, eds., *Das Ende der Parteien* (Düsseldorf, 1960), p. 119n.

6. Walter Struve, *Elites against Democracy: Leadership Ideals in Bourgeois Political Thought in Germany, 1890–1933* (Princeton, N.J., 1973), pp. 363–64; David Schoenbaum, *Hitler's Social Revolution: Class and Status in Nazi Germany* (New York, 1967), p. 7.

7. Matthias and Morsey, eds., *Das Ende der Parteien*, pp. 240, 239.

8. Schoenbaum, *Hitler's Social Revolution,* p. 8; Jürgen Kocka, "Zur Problematik der deutschen Angestellten 1914–1933," in Hans Mommsen et al., eds., *Industrielles System und politische Entwicklung in der Weimarer Republik* (Düsseldorf, 1974), pp. 802–809. See also Sandra J. Coyner,

"Class Consciousness and Consumption: The New Middle Class during the Weimar Republic," *Journal of Social History,* vol. 10, no. 3 (1977):310-332. As Coyner notes, analysts have tended to describe an objective employment situation and to infer the kind of consciousness that seems appropriate by their own standards. Coyner also suggests that there are significant differences in outlook between salaried employees in the private and public sectors, but there is no need to discuss that here.

9. See the discussion in Adolf Grünbaum, "Free Will and Laws of Human Behavior," *Psychoanalysis and Contemporary Science,* vol. 3 (1974): 5.

10. Sigmund Freud, "Inhibitions, Symptoms and Anxiety," in James Strachey, trans. and ed., *The Standard Edition of the Complete Works of Sigmund Freud* 24 vols. (London, 1956-1961), 20:77-174.

11. Arthur H. Schmale, Jr., "Depression as Affect, Character Style, and Symptom Formation," *Psychoanalysis and Contemporary Science,* vol. 1 (1972):323-351; Schmale, "A Genetic View of Affects: With Special Reference to the Genesis of Helplessness and Hopelessness," *Psychoanalytic Study of the Child,* vol. 19 (1964):287-310. T. L. Dorpat, "Depressive Affect," *ibid.,* vol. 32 (1977):13-17. On the signal function of guilt, for example, see Anton V. Kris, "On Wanting Too Much: The 'Exception' Revisited," *International Journal of Psychoanalysis,* vol. 57, nos. 1, 2 (1976):92; on the signal function of humiliation, Julian L. Stamm, "The Meaning of Humiliation," *International Review of Psychoanalysis,* vol. 5 (1978):425-433.

12. George S. Klein, *Psychoanalytic Theory* (New York, 1976), pp. 131-134.

13. W. G. Joffe and Joseph Sandler, "Comments on the Psychoanalytic Psychology of Adaptation with Special Reference to the Role of Affect and the Representational World," *International Journal of Psychoanalysis,* vol. 49 (1968):450-451; W. G. Joffe and Joseph Sandler, "Notes on Pain, Depression and Individuation," *The Psychoanalytic Study of the Child,* vol. 20 (1965):395-422; Joseph Sandler, "The Background of Safety," *International Journal of Psychoanalysis,* vol. 41 (1960):352-358; Joseph Sandler and W. G. Joffe, "Psychological Conflict and the Structural Model," *ibid.,* vol. 50 (1969):79-90; Karl H. Pribram, "Emotion: Steps towards a Neuropsychological Theory," in David C. Glass, ed., *Neurophysiology and Emotion* (New York, 1967), pp. 3-40. Affect had become in Freud's work a communicative form in the service of adaptation as opposed to being an indicator of discharge, which it was earlier. Psychoanalysts have by now established more than one theory of motivation, and as John E. Gedo has pointed out, they may yet end up with none. The introduction of affect theory is meant to resolve this problem at least for psychoanalytic sociology. See Gedo, "Theories of Object Relations: A Metapsychological Assessment," *Journal of the American Psychoanalytic Association,* vol. 27, no. 2 (1979):367-368.

14. Joseph Sandler, "Dreams, Unconscious Fantasies and 'Identity of

Perception,'" *International Review of Psychoanalysis,* vol. 3. no. 1 (1976): 36. "It is worth noting again that the 'force' behind the wishful fantasy . . . need not be an instinctual drive but can equally be the need to avoid painful experiences and preserve safety and well-being."

15. Arthur H. Schmale, Jr., Reporter, "The Sensory Deprivations: An Approach to the Study of Affects," *Journal of the American Psychoanalytic Association,* vol. 22, 4 (1974):636.

16. Joseph Sandler and W. G. Joffe, "Towards a Basic Psychoanalytic Model," *International Journal of Psychoanalysis,* vol. 49 (1968) :79–90

17. George Klein, "The Vital Pleasures," *Psychoanalysis and Contemporary Science,* vol. 1 (1972) :181–205: In other words, what begins as an undifferentiated reaction to need and gratification becomes, through development in contact with others, a differentiated network expressive of various kinds of anticipated pleasures on the one hand and dangers and deprivations on the other.

18. C. Janet Newman, "He Can but He Won't," *Psychoanalytic Study of the Child,* vol. 28 (1973) :87.

19. Joffe and Sandler, "Comments on the Psychoanalytic Psychology of Adaptation," pp. 450–451.

20. Pribram, "Emotion: Steps towards a Neuropsychological Theory," pp. 3–40.

21. George Klein, "On Hearing One's Own Voice," in Max Schur, ed., *Drives, Affects, Behaviors,* 2 vols. (New York, 1965) :90. The reference here is to anxiety, but obviously the expression of affect will vary from person to person, making it necessary to consider the widest possible variety of affective responses to events. The point is that one affective expression or another will appear on every occasion that the conditions assuring protection or gratification (of interest, morality, and/or wishfulness) cannot be met. Moreover, we may on observational grounds consider that ego capacities have varied sources, and becoming relatively independent of drives, are narcissistically cathected as expressions of character. Thus, the sense of adequacy and efficacy in adults functioning within economic and political spheres is based on these relatively nonsensual, sublimated, desexualized interests. Insofar as familial and religious commitments are affected, the situation would have to be described in other (sexualized or sensual) terms. But threats to self-control, efficacy, the ability to act on culturally valued skills, and so forth, can inspire feelings of ruthlessness and a great capacity for violence.

22. The Nazis won 18.3% of the vote in the Reichstag elections of September 1930 (6,410,000 votes, 107 seats in the Reichstag, compared to 810,000 votes and 12 seats in 1928).

23. On unemployment figures during the depression, Dieter Petzina, "Germany and the Great Depression," *Journal of Contemporary History,* vol. 4, no. 4 (1969) :60; Ernst Deuerlein, ed., *Der Aufstieg der NSDAP in*

Augenzeugenberichten (Düsseldorf, 1968), pp. 392–394; Albert Wucher, ed., *Die Fahne hoch: Das Ende der Republik und Hitlers Machtübernahme* (Munich, 1963), pp. 9–13; Axel Eggebrecht, *Volk ans Gewehr: Chronik eines Berliner Hauses 1930–1934* (Frankfurt a/M., 1959), p. 8. This is a fictionalized account of events but very much to the point.

24. Ernst Robert Curtius, quoted in Deuerlein, *Der Aufstieg*, p. 369.

25. Harry Kessler, *In the Twenties: The Diaries of Harry Kessler*, trans. Charles Kessler (New York, 1971), p. 408.

26. Leopold Schwarzschild, *Die letzten Jahre vor Hitler: Aus dem 'Tagebuch' 1929–1933* (Hamburg, 1966), pp. 164, 224.

27. Goebbels, quoted in Viktor Reimann, *The Man Who Created Hitler,* trans. Stephen Wendt (London, 1976), p. 155; Walther Hofer, *Der Nationalsozialismus: Dokumente 1933–1945* (Frankfurt a/M., 1957), pp. 24–25.

28. Deuerlein, *Der Aufstieg*, p. 407; Wolfgang Horn, *Führer-ideologie und Partei-Organisation in der NSDAP 1919–1933* (Düsseldorf, 1972), pp. 349, 352, 356, 362–63, 370, 376; Jeremy Noakes, *The NSDAP in Lower Saxony, 1921–1933: A Study of National Socialist Organization and Propaganda* (Oxford, 1971), pp. 233–34.

29. Werner Jochmann, ed., *Nationalsozialismus und Revolution: Ursprung und Geschichte der NSDAP in Hamburg 1922–1933. Dokumente* (Frankfurt a/M., 1963), pp. 419–420. See also *ibid.,* p. 416, on the problem of cathectic instability in the period November 1932, as people struggled for control.

30. Erich Kästner, quoted in Kathcrine Larson Roper, "Images of German Youth in Weimar Novels," *Journal of Contemporary History*, vol. 13 (1978):508–509; Heisenberg, quoted in Alan D. Beyerchen, *Scientists under Hitler: Politics and the Physics Community in the Third Reich* (New Haven, Conn., 1977), p. 58; and Joseph Haberer, *Politics and the Community of Science* (New York, 1969), p. 164. See also the statement by Reinhold Schneider, in Deuerlein, *Der Aufstieg*, p. 411; Oskar Maria Graf, *Gelächter von aussen: Aus meinem Leben 1918–1933* (Munich, 1966), pp. 461–471, 490–492, 506–512; Max Tau, *Trotz allem!* (Hamburg, 1972), pp. 119–123, 126–128; Egon Larsen, *Weimar Eyewitness* (London, 1976), pp. 150–152, 162–163; Otto Reinemann, *Carried Away* (Philadelphia, 1976) pp. 68–70.

31. Quoted in Michael Stephen Steinberg, *Sabers and Brown Shirts: The German Students' Path to National Socialism, 1918–1935* (Chicago, 1977), p. 8.

32. Peter D. Stachura, *Nazi Youth and the Weimar Republic* (Santa Barbara, Calif., 1975), p. 45. Michael Kater, *Studentenschaft und Rechtsradikalismus in Deutschland 1918–1933* (Hamburg, 1975), pp. 11, 116–117.

33. Matthias and Morsey, eds., *Das Ende der Parteien*, p. 215.

34. Jochen Klepper, *Unter dem Schatten deiner Flügel: Aus den*

Tagebüchern der Jahre 1932–1942 (Stuttgart, 1956), p. 42. See also Klepper, *Briefwechsel 1925–1942*, Ernst G. Riemschneider, ed. (Stuttgart, 1973), pp. 227–229. See Jonas Lesser's comments on Klepper's diaries, *The Wiener Library Bulletin*, vol. 11, no. 1, 2 (1957):7.

35. Jochmann, ed., *Nationalsozialismus und Revolution*, p. 421.
36. Hans Bernd Gisevius, *To The Bitter End* (Cambridge, Mass., 1947), pp. 93–95. Hans Carossa described his meeting on March 5, 1933, in Munich with a young pharmacist's assistant who had helped to destroy a Social Democratic newspaper office. The youth told Carossa it was the happiest night of his life. Hans Carossa, *Ungleiche Welten* (Hamburg, 1959), p. 25.
37. Hildegard Brenner, *Die Kunstpolitik des Nationalsozialismus* (Hamburg, 1963), p. 188. Paul E. Kahle, Naumann's colleague at Bonn and an anti-Nazi who subsequently fled, described Naumann as a great idealist, an honest man, and an excellent teacher. Paul E. Kahle, *Die Bonner Universität vor und während der Nazi-Zeit (1923–1939)*, Wiener Library document, n.d., p. 5, on Naumann and his talk.
38. Kahle, *Die Bonner Universität*, p. 6.
39. Guido Schneeberger, *Nachlese zu Heidegger: Dokumente zu seinem Leben und Denken* (Bern, 1962), p. 71.
40. Josef Weiss, quoted in Hans Peter Bleuel and Ernst Klinnert, *Deutsche Studenten auf dem Weg ins Dritte Reich: Ideologien, Programme, Aktionen 1918–1933* (Gütersloh, 1967), p. 244.
41. Professor Julius Petersen, Berlin, quoted in Hans Peter Bleuel, *Deutschlands Bekenner: Professoren zwischen Kaiserreich und Diktatur* (Bern, 1968), pp. 213–214.
42. Alfons Ilg, quoted in Josef Wulf, *Literatur und Dichtung im Dritten Reich: Eine Dokumentation* (Gütersloh, 1963), p. 55.
43. W. Harless, quoted in George Mosse, *The Crisis of German Ideology* (New York, 1964), p. 24.
44. Emil Hirsch, quoted in Bleuel, *Deutschlands Bekenner*, pp. 223–24. See also Wendula Dahle, *Der Einsatz einer Wissenschaft* (Bonn, 1969), pp. 180, 185, 186, 219, 232.
45. Karl Kraus, *Die dritte Walpurgisnacht* (Munich, 1952), pp. 15, 254.
46. On the Third Reich as promising a new form of being, the breakthrough to a new form of life, a new experience of reality, and so forth, see Dahle, *Der Einsatz einer Wissenschaft*, pp. 180, 185, 224. On Hitler and Christ, for example, Walther Hofer, *Der Nationalsozialismus: Dokumente*, p. 128; or Robert G. L. Waite, *The Psychopathic God* (New York, 1977) pp. 31, 82–83. On Nazism as a Copernican event, see Josef Ackermann, *Heinrich Himmler als Ideologe* (Göttingen, 1970), p. 108.
47. The affective experiences that follow from the inability (or feared inability) to produce familiar behaviors are further deepened by the implication of drive expression, as sublimated forms of wishful gratification are

also affected. Once normal routines are disrupted, it becomes not only a matter of interpreting the import of adverse or challenging events, it becomes also a matter of controlling one's feelings about the ability to master events, and this is obviously rendered more difficult by the continuing need to integrate affective, drive, and environmental threats. Thus, persistent environmental dislocation will not only undermine hitherto socially adaptive capacities, it will also lead to more global and undifferentiated responses. At this level, even unchecked, commonsense guesses cannot be effectively managed because the threat appears so severe that it is no longer possible to relate perceived elements of the threat in common sense terms, but only in terms of the effects of damaging events on the self. This may nevertheless appear "syntonic" because disruptive experience can be absorbed in thought and action only as people are currently able to evaluate their situation, that is, in ways that seem reasonable and logical under the circumstances. This level of affective expression may well not have been held constant over time as different events affected the interests, moral commitments, and wishful expectations of individuals and groups differently. Workers could withdraw their efficiency, Catholics could protest interference with their beliefs, and any one of these individuals quoted could have changed his mind, as we saw in the first chapter. But the Nazis did evoke a great deal of enthusiasm, especially in the years 1933–1939, and they could always count on a high degree of public loyalty after that, even if people were not enthusiastic about the course of the war. The morale of some Nazis remained high, and the Nazis were able to recruit, among the young, for example, to the very end.

48. The process by which Nazism penetrated the society down to the lowest levels is well described by George Mosse, *Nazi Culture* (New York, 1966), pp. 275, 375–77, 383. On the matter of opportunism, Karl Thalheim wrote in 1941 that "with a total of far more than ten million Jews in Europe, the question of the future fate of this great mass is of the utmost importance. The situation is quite different now from what it was in the past; assimilation is no longer the goal, but separation, a complete break between irreconcilable racial elements. . . . [T]his trend towards separation must lead us to expect in the future further mass movements of population, all with the purpose of the inner pacification of Europe." Then, in retrospect, Thalheim wrote, "I never believed in the appalling nonsense of these sentences. I was never anti-Semitic, not even in the Nazi era; before 1933, to a certain degree far into the Nazi period and after 1945, I was and am a close friend of many people of Jewish origin. People will say: 'All the worse for you'—and they will be right. Anyone who wrote such things and believed them to be true is less guilty than one who wrote them without believing them. I now know—and not just as of today—that this was my real betrayal of spiritual integrity." Rolf Seeliger, ed., *Braune Universität: Deutsche Hochschullehrer gestern und heute, 6 vols.* (Munich, 1966), 4:32, 36. See also

the statements of Gerhardt Giese, a student of Eduard Spranger's, *ibid.*, p. 115.

49. Sigmund Freud, *Totem and Taboo,* trans. James Strachey (New York, 1950), 95.

50. Note Erikson's definition of psychoanalytic sociology in "Identity and the Life Cycle," *Psychological Issues,* vol. 1, no. 1 (1959):151.

51. Freud was certainly aware of the impact of potentially traumatic events and often addressed himself to the problem. But neither the effects of traumatic events nor the effects of systematic, legitimated social change were ever properly integrated in theory.

52. The emphasis here must be on the feelings evoked by the immediate situation and on the capacity of contemporary organizations to help control the feelings, and not, for example, on historically or culturally available ideas. The ideas of blood, soil, folk, nation, expansionism, revanche, military prowess, cultural or racial superiority, etc., fed into and lent structure, coherence, and a sense of continuity in a situation in which people were threatened by the loss of control over personal, social, and national destiny, a result of real environmental failure, the last and potentially most damaging in a series of failures reaching back to the defeat in the Great War. Those elements from the historical past that the Nazis used for their own ends may have been dramatic and available, but, as with racism, they were not always the most important by any means. It was the effects of the unique, immediate situation that allowed culturally familiar expressions to be used as if the meanings were constant or everyone meant the same thing by them.

3

Hitler's Charismatic Power and Racism

The integration of individuals and groups in a unified Reich after the Nazi seizure of power was meant to include those formerly opposed to Nazism because of class and religious loyalties and to restore a sense of the capacity for effective action through the society. The process occurred more quickly and with less conflict than anyone anticipated, but it was not entirely without conflict, particularly as radical elements in the SA pushed to increase the pace of social change. This seems paradoxical, but it happened in this situation as it happened in other revolutionary situations, because the form and extent of change, when it actually unfolds, is always different from the interests, moral commitments, and wishful expectations that prompted participation in the first place. The shape of events can never be anticipated, the bases for making decisions about events are never entirely shared or constant, and rapid, violent change always has the sense of being forced change, even for otherwise loyal individuals and groups, because it seems to require finally the unexpected and unwelcome renunciation of important positions. This accounts for the persistent discontent among more radical elements of the SA leadership and rank and file, who objected particularly to Hitler's rejection of a "second," social, revolution.

Hitler could not afford the social ambitions of the SA, either with respect to the military or the economy. Hitler needed the support of the army and industry, especially over the short run, in order to realize his own ambitions for the revolution, which were much more racial than social. He therefore moved to destroy elements of the SA leadership and to weaken the power of the organization. Ernst Röhm responded by repeating the well-worn observation that revolutions devour their own children. Indeed, revolutions must devour at least some of their own as any practical, far-reaching decision is bound to affect how people feel about the world, fostering an unanticipated but definite sense of violation and, consequently, a spirit of opposition.[1]

Hence, the variety of popular response after the Nazis came to power included great expectations for profound change, exaggerated moral expression, a sense of recovery of the past among some, a sense of release from the past among others, but included as well overt hostility and ambivalent wariness, a result of thwarted expectations. The process of integration then had to encompass all these different responses, the Nazis had to hold people together and provide a shared background against which such events as the massacre of June 30, 1934, could appear legitimate. The success of the Nazis in this effort depended upon two subjective but sociologically relevant factors, leadership and ideology.

Leadership and ideology were decisive for the fate of the movement all along. They were so initially for many people because there were no consciously mastered socially available techniques for dealing with the crisis of the early 1930s and subsequently because the regime needed to provide a background for action, a consistent, integrated orientation to past, present, and future. Hitler and the Nazis did manage to constrain a variety of threatening experiences, lending a degree of structure to the field, answering to unwelcome emotional experiences. The Nazis succeeded in directing attention to the social world, promising, and by their own actions giving evidence of, an uninterrupted capacity to master that world.

The immediate problem is to explain the basis for Hitler's appeal in this situation: to identify the process by which Hitler was able to substitute for the internal conflict provoked by the crisis an interpersonal tie with himself as the representative of order, morality, and change, restoring an active sense of purpose to a large number of people. Weber's concept of charismatic authority, focused on the impact of disruptive situations, provides some sense of the nature of that appeal, which was based on the personal capacity of an extraordinary individual to make sense of a world in which routine sequences of thought had been broken up, making it difficult to interpret events by familiar criteria, in which the capacity to remember had been adversely affected (memory being the way people have of confirming the sense that there is a way out, as there had been before), and in which the connection between the inner world of mental representations of self and others and the everyday social world had become uncertain and problematic.*

The immediate situation in Germany, the subjective sense of the Weimar Republic as weak and failed, even treasonous and evil, widespread after 1930, may certainly be characterized in these terms. This situation did not create Hitler, as he came to it with considerable political and oratorical talents of his own. Rather, it facilitated what Ernst Deuerlein has referred to as "Hitler's Ermöglichung," that is, it made Hitler possible, providing him with a milieu for the exercise of those talents.²

For some Germans, obviously, the sense of weakness and failure occurred quite early, a result of the defeat, the Versailles Treaty (especially the war guilt clause and reparations), the parliamentary form of government, persistent border conflicts (as

---

* Weber's concept of charismatic leadership provides no sense of the internal dynamics of charismatic leaders, nor does it help to identify the dynamic links between leader and led. However, charismatic leaders as a sociological phenomenon are defined not by shared dynamics or styles nor by ideological commitments, which vary. In a sociological sense, charismatic leaders are defined by a shared capacity for ordering social situations in which the ability to act on culturally valued ideals and character traits appears threatened or disrupted. The personal dynamics of a charismatic leader and the dynamic links between the leader and the led must be specified anew in each situation in which such a leader appears.

Germany seemed constantly threatened with further losses), economic pressures, and other reasons as well. But the sense of failure spread rapidly and widely through the population after 1930, as society, or the various authorities representing society, were no longer perceived with any degree of confidence as able to sustain everyday life as people wanted to live it. By contrast, Hitler's capacities as a leader allowed activity to become more organized for many people, as he also worked to strengthen the view individuals and groups had of themselves in a future related to a psychologically useful past and present.

Hitler displayed considerable skill as a political leader, employing a number of consciously developed techniques for mastering the situation by manipulating his audience. His posture as an anonymous soldier, a humble man risen from obscurity, for example, was meant to indicate that he was like them and, more important, that they were like him, that they could share his grandeur in lesser or greater ways through activity of their own. They were all part of the same community (blood), though for political reasons he had to appear elevated and superior. Hitler often said that he could fathom the innermost concerns of the German people, and in many respects he could. Hitler promised the Germans he would restore them to a special and unique place, holding himself forth as an example of willful effort in defiance of the contemporary reality. And people who know the pain of reality are prone to admire those who appear defiant of it.

Hitler tried to restore an ideological conception of the world sufficiently stable for people to experience society again as a safe, supportive, enhancing place, and he was quite deliberate in the way he appeared to people and in what he told them. Thus, in his proclamation to the German nation, February 1, 1933, Hitler spoke of unity, energy, vitality, discipline, honor, sacrifice, the legitimate suppression of enemies, the termination of a chaotic era, the recovery of imperishable virtues that had given rise earlier to imperishable accomplishments.

> The National Government will regard it as its first and foremost duty to revive in the nation the spirit of unity and cooperation. It will preserve and defend those basic principles on which our

nation has been built. It regards Christianity as the foundation of our national morality and the family as the basis of national life.³

Hitler did not believe that Christianity was the foundation of morality, nor did he necessarily believe in the primacy of the family or of national as opposed to racial life; but that was of little consequence, after all.

It is necessary to stress that Hitler's success depended as much on what he told people as it did on how he appeared to them. It was not as if any content would have served, just because he was, or appeared to be, a powerful and able individual. On the contrary, the content had to be specific to the concerns of the people, and the language, the ideological conception of the world he offered, was important because it was the overt public basis on which to reassure people that a safe, supportive society could be restored, it was a way of linking people to each other, directing them to socially useful activities, despite the idiosyncratic nature of their perceptions and the heterogeneity of their strivings.⁴

Hitler's style, his public image, and his ideological expression are connected in a particularly obvious way. The style was based on an intensity of conviction constantly communicated in highly emotional speeches characterized by the threat of violence and by an exalted vision of what Germany must become. It was an important part of Hitler's public style never to appear weak or to show fear, never to supplicate or appear to be the passive victim of events, never to accept a subordinate position. Hitler then used this style to impress on people a particular ideological content comprised of four principal themes and ideas: that the past was sacred, that struggle was inevitable and desirable, that will, intuition, and instinct were of a higher order than reason, and that hierarchy was a valid principle that must be enforced. But he was vague enough, or even misleading enough, for people to project whatever meaning was important for them into this content. Hitler lied to people obviously, but he was also more subtle than that: He often did not mean what he knew people would infer from his statements because he shared a language with them.

In fact, the connection between public style and ideological content was fused in Hitler's mind by the single idea of racial anti-Semitism, which served to elevate him, prohibit any solutions other than the ones he proposed, and identify at once the source of the peril to the community and the reason that compelled its spiritual mobilization. People knew that Hitler was anti-Semitic, but bound by their common sense conceptions of the world, most of them did not understand the force with which he was attached to this idea or how it made him different from them.

Hitler's radical form of racial anti-Semitism was obviously present in Austro-German culture, but it emerged in his life story in a revealing way and it acquired a singular force that merits close attention. Take for example the threads of symbolic paradox, the reconciliation of opposites, that run through Hitler's autobiographical statements. The paradoxical image of the anonymous soldier–world historical leader, already referred to, was one; the paradoxical image of the wise child was another. Hitler explained in *Mein Kampf* that even as a child his oratorical talent was being developed, school work was ridiculously easy for him, he learned to understand and grasp the meaning of history, he knew the difference between dynastic patriotism and volkish nationalism, he had learned that Austria must be destroyed so that Germany could be protected, and he had already become an artistic and political revolutionary.[5]

The anonymous–famous paradox was meant to underscore Hitler's successful defiance of social convention, to show that even the greatest barriers could be surmounted if one but had the will. The wise child paradox was meant to show that all this was nevertheless internally consistent with the course of his life. However, Hitler's childhood wisdom did not encompass anti-Semitism. Hitler explained in *Mein Kampf* that anti-Semitism did not come early or easily; it came to him rather as a young man in Vienna, at the cost of great inner struggle that culminated in the most significant transformation of his life. Reason won out over sentiment only after months of struggle, and even then his thinking faltered for weeks at a time, once even for months. Hitler was tormented by the feeling that perhaps he was being unjust, a feeling that made him anxious and uncertain. "For me this was the time of the greatest spiritual upheaval I have ever had to go through"—fearful and oppressive thoughts came to him "in profound anguish."[6]

These autobiographical statements of Hitler's may be taken as true, not in their factual context, since his statements concerning time and place are contradicted by evidence, but as the description of a real emotional experience.[7] Hitler was not describing a historical event in any conventional sense, he was describing what he was, what he wanted to be, and what he did become. The truth value of these statements is in his feelings about the experience, not in his ability to say where or when it actually happened. Whether Hitler wanted people to think that he had arrived at his racist conclusions long before he ever really did or whether this was how he actually remembered it is an interesting but not a pressing question.

What matters is that Hitler became an extraordinarily effective leader in terms of the experience he described. For out of the resolution of this mental anguish, whenever and wherever it occurred, Hitler emerged with a fixed and irrevocable image of the Jew as a monster, a pestilence worse than death, striving for the collapse of civilization and the devastation of the world. Hitler had been taken in, he had experienced a rude awakening, and once he worked it through the scales fell from his eyes and a thousand things that he had never seen or understood before became clear to him. The world became coherent at last, the terms of the resolved conflict being the very terms that Hitler constantly invoked as truly German or Aryan, the opposite of everything weak and vile: He had wrested clarity from confusion, certainty from ambiguity; he would never again be the passive object of events or a soft, sentimental victim; he would be the active master of events, the hard witness and critic of reality. Hitler had gone through a bitter inner conflict; he resolved the conflict by locating its source in the social world, raising himself from a spirit of futility to a spirit of longing and conviction, from a sense of weakness to a sense of strength, bringing order to his perceptions by pseudoscientific and semiscientific cognitive procedures, linking himself to that social world by his understanding of "the eternal laws of nature." Hitler thus acquired the psychic means, which he always thought of as the brutal determination, to carry on the necessary racial struggle.

Hitler described other insights and conclusions in these pages of *Mein Kampf,* especially his awareness of the ease with which the Jews could change their shape. Hitler reported his reaction to the Orthodox Jews who appeared in traditional garb in Vienna, so unlike the assimilated Jews he had seen in Linz. The Jews in Linz had spoken German and had otherwise behaved like Germans. But this was mere deception: The essence of Jewry was revealed by these other ones, the visible, physical representatives of Jewish nature. Hitler concluded that all Jews were intrinsically, unalterably alike and all distinctions between the assimilated and the traditionalists, or between Zionists and liberals, which they themselves busily peddled, were false.[8] It was in this sense that Hitler once said that Julius Streicher idealized the Jews and never showed how diabolical and cunning they really were. He meant that Streicher's pornographic gutter displays showed the Jews only as greedy, physically repulsive, and sexually perverted. But any Jew could easily have appeared otherwise and it was the ease with which they could appear one way and think another that made them dangerous.[9]

In this world Hitler created of literal, concrete images, two people could not occupy the same space. Aryans and Jews both made claims to be exceptional, to be chosen, but it could only be true of one of them. The Aryan claim was based on the observation of nature and reality, and its source was perfectly clear. The Jewish claim, however, originated from their false, divided, hidden nature, their inner chaos, which compelled them to foist on the world such despicable, leveling creeds as Christianity and democracy, in order to create a world true to their own degenerate spirit.

The conclusion was obvious: Jews could appear to be like but were in fact the opposite of Germans or Aryans. No Jew could ever represent or stand in for him, for example. They were not his double, they were his opposite, and the lower they fell the higher he rose, a height ultimately acknowledged by their envious and malicious attacks on him. From this followed still another conclusion: The more he came to hate these Jews, the more he loved the Germans, who were only their unwitting victims. And this was the trinity that shaped all subsequent action for him: Hitler, the Jew, and the Victim.

There were undoubtedly many reasons why Hitler should

have experienced grievous doubt and turmoil before he could accept these conclusions. But one source in particular is rather easily identified: Hitler *knew* Jews who had been kind to him or had done him favors,[10] and he knew enough about the Austro-German culture to know that there were Jews who were creative and courageous by any standard. Hitler could not appear crazy to himself, and still he had to credit his senses. Hitler resolved this turmoil by accepting a conclusion already available, that the Jews are a *principle,* that Jews as a group are a specific, harmful racial and social phenomenon. What any individual Jew did was then of no consequence because idiosyncratic instances do not count.

The kind of inner struggle and resolution that Hitler described may or may not meet the opposition of others. But it does meet the prohibitions of reality as expressed in the laws of language, logic, and causality. However, once Hitler defined the Jews as a racial principle he no longer had to appear confused to himself and he did not care how he appeared to others. He was prepared to meet all opposition and risk death if it came to that, because the prospect of physical extinction did not frighten him nearly as much as the prospect of psychological extinction, which he had once clearly faced.

The position Hitler had arrived at is formulated in his first extant statement of political anti-Semitism, of September 16, 1919. This statement warrants close examination because Hitler had managed to generalize the result of his transformation and was able to offer it as a strategy and a world view. Hitler explained that the sense of danger Germans feel in the presence of Jews is expressed in an evident aversion to them. This aversion results, however, from the personal impression individual Jews make and not from any awareness of their systematically destructive nature.

> Thus, anti-semitism all too easily takes on the character of a mere manifestation of emotion. And that is wrong. As a political movement, anti-semitism cannot and must not be determined by emotional motives but by a recognition of the facts . . . to begin with, Jewry is incontestably a race and not a religious community.[11]

Hitler's distinctions were obviously important: People could change their religion, but race was immutable. And they were accurate at least to the extent that it made a difference in Germany whether one was anti-Semitic because of what Jews did, if Jews were also seen as part of an abstract social process that affected them as it might have affected others, or whether one was anti-Semitic because of what Jews are "in nature," or as the racists put it, "in the blood," referring to heredity, genetics, biology, race, the body, in any case referring to something nonabstract, literal or concrete.

In the former instance, East European Jews may have been despised as rude, uncivilized, shamefully different, and in constant violation of German standards for good taste and appropriate behavior. But baptized, educated, assimilated Jews were acceptable enough: They behaved like Germans, they were capable of changing and therefore assimilable in the larger community.[12] In the latter instance, however, Jews were perceived as bad in themselves, they were not capable of changing, and they were not assimilable in the larger community. The Jews were a "foreign race," neither willing nor able to sacrifice their racial characteristics or to renounce their own way of feeling, thinking, and striving.[13]

The significance of this focus on what Jews are, as opposed to what they might do, always expressed in the most concrete body terms, that is, initially as metaphor but finally in a very concrete way, may be hard to grasp for anyone who thinks metaphorically without reflecting on the process.* But take the phrases "all

---

* Religious people may insist on the literal rather than on the abstract, metaphoric, or symbolic meaning of one or another ritual belief or practice. They may be able to define the difference, choosing to accept the literal meaning, or they may be unable to accept standards by which to define the difference, as a function of "internalized morality." I refer here to "concrete" meaning in addition to "literal" meaning because Hitler, at least as far as Jews were concerned, was incapable of understanding or accepting these different levels of meaning, as a result of a cognitive lapse or incapacity and not as a result of internalized moral commitments to a venerable, received tradition. Technically speaking, the terms "metaphoric," "literal," and "concrete" as discussed here represent three different levels of metaphoric thinking. It is difficult to explain the problem more precisely be-

or nothing at all" and "I've got you under my skin." These are metaphoric expressions of emotional experience, as the phrase "it was a slap in the face" is a metaphoric expression of an emotional experience. When an impassioned lover says "all or nothing at all," he or she does not really mean nothing; typically, the other's life is not at stake as people have learned to settle for more modest things and even to absorb disappointment. When a person enamored or hateful of another says "you're getting under my skin," the metaphoric reference is to an emotional, not a physical penetration. However, when Hitler talked about the Jewish virus, or when he described the Jews as parasites on the body of other people or on the body politic, he meant precisely that; the Jews were not *like* parasites, they *were* parasites. Hitler meant that Jews caused tuberculosis and syphilis, they drained valuable substance, they corrupted or poisoned the blood of others while preserving their own, they shamelessly and unscrupulously victimized people from whom they literally stole the inner substance, who they were sucking dry and weakening with the prospect of enslavement. The only thing he was not sure about was whether the Jews did these things consciously or unconsciously, though in either case the impulse to dominate was racial; Jews were driven by their blood and most of them probably did not know their own power.[14]

By the time that Hitler's body language moved from metaphoric to concrete expression, as it appears, for example, in *Mein Kampf*,[15] he was no longer able to distinguish being from doing so that failure or defeat always rendered problematic the possibility of continued existence. And when he said that Germany will either be National Socialist or nothing, or the Aryans will

---

cause neither Piaget's conception of "decentering" nor Freud's topographical conception of "preconscious thinking" is complex or differentiated enough to encompass it more precisely. However, if biographical comments on the younger Hitler can be taken seriously, then Hitler had at an earlier point a capacity for higher level metaphoric thinking that he subsequently lost. I will therefore discuss Hitler's cognitive incapacity in the terms "metaphoric" and "concrete" for the sake of making a necessary distinction, without implying that in a technical sense this is refined enough to be "correct."

either triumph over the Jews or mankind would descend into a dark void from which it would never return, he meant exactly that. Hitler acted with ruthless intensity because he conceived of his struggle as so unique and all-encompassing, and he was so literal about "the blood," that there could be no chance of recovery in the event of failure, no hope of having learned anything so that one could do it better the next time. Of course, Hitler did not want his genius to go unrecognized, nor did he want to disappear without a trace, and faced with the gloomy prospect of defeat, he would say that perhaps in a hundred years the blood would be sufficiently replenished for people to appraise correctly and hence to cherish his contribution. In the end, however, he was convinced that the future belonged to the stronger peoples of the East and he ordered the destruction of Germany because he felt that the Germans deserved to perish.

Hitler's language and thought as a racist reveal a level of cognitive capacity discussed by Piaget in terms of a transition people routinely make from "egocentric" thinking and a naive "absolute realism" to a higher level of thinking that recognizes the relativity of human qualities, a recognition dependent upon discovering the relativity of self. Hitler undoubtedly had this capacity once, but he lost it, and he could not relativize human qualities in this sphere of thought, he could not think about Jews except as a degenerate, contaminated racial group devoted to the destruction of Aryan blood, a goal he came to believe he was providentially chosen to block.

Hitler said many times that once this fundamental view of the world was formed he never had to change anything.[16] What he meant was that his feelings about the world, his subjective experience of it, never had to change. Of course, Hitler learned things, reoriented himself to different exigencies and so forth. Still, this basic sense of the world remained untouched and untouchable. The trinity—Hitler, the Jew, and the Victim—was so vivid and real for him that it lost the quality of content and became—to be as concrete as he was himself—a part of the structure. There was no multiplicity of meanings in complex actions, no alternative possibility to the roles he assigned to each, there was no relativity, and especially not for Hitler and the Jew, because if the Jew was not what Hitler said he was, then Hitler was not at all. He therefore clung to that figure down to the last days in the bunker, his final message an admonition to pursue the struggle against the world poisoner of peoples. Hitler preferred to die with his mind intact, fighting to maintain his omnipotent grandeur, rather than question any fraction of what he had done.

These racist conclusions enabled Hitler to assimilate visions of good and evil, represented by Aryans and Jews, to one coherent conception of struggle. This was for Hitler a breakthrough *to* reality, it was the key to his visions and to his greatness. The Jew in his mind was the mightiest counterpart of the Aryan, as destructive and chaotic by nature as the Aryan was creative and disciplined. The Jewish form was unstable because the Jewish soul was fractured; Aryan form was stable because Aryan soul was pure, or could be purified through breeding and training. Hitler knew there were blond, blue-eyed Jews, but they could never be assimilated because in a generation or two they would revert to type. Hitler knew there were Jews who "behaved correctly" towards Germany, but that did not matter because they never defended Germany against Jews who did not behave correctly.[17]

Once Hitler fastened upon his racist truth, his attention narrowed and focused on it, he repeated it continuously and elaborated upon it endlessly. The truth elevated him and degraded them, confirmed his virtue and their depravity, established him as Aryan, the Aryans as the creators of everything valuable, and established them as a sham people. All weakness, everything shameful and humiliating was banished from him, as was every evil inclination. And in a generous act of complicity it was all shared out among Germans and Jews, while he stood uniquely above the lot, free to act to the greatest extent because he was acting from urgency, purity of motive and insight.

Thus, in his initial statement of September 1919, Hitler observed that the mere expression of anger over what Jewish individuals and groups did led only to spasmodic violence, to emotionally inspired pogroms which passed, allowing the Jews to carry on as before. Germany had to move on to the anti-Semitism of reason, the Jews had to be approached in the light of science and the solution had to be systematic because the problem was systematic. People simply had to be cold and determined about it, and the final goal, expressed with an inevitable degree of ambiguity, "must always remain the removal of the Jews as a whole." [18]

Hitler's world view was therefore based on a thoroughly integrated conception of ideal images and ideal ambitions, perfect form, life in the body, a life of action and brutal struggle equal to the truth of nature that any sober observer could see. Hitler insisted that people were awed by ferocity, that they would behave ferociously themselves if they dared, and more were ready to dare than anyone imagined. People needed encouragement and a sense of rectitude, which he would provide. The point was to strip away a civilized layer of feeling, sentiment, concern, an unreflected accretion of civilized muck that would reveal the natural man as he was before the Jews tried to destroy him, initially through Christianity and finally through Bolshevism. This natural man was instinctively aware of and devoted to blood, community, hierarchy, discipline, and sacrifice, a merciless fighter in the service of these principles.

This was what Hitler meant to restore or recover by his activities. The young, he told Hermann Rauschning, must be violently active, indifferent to pain, they must exhibit the gleam of pride, the independence of the beast of prey. He would eradicate thousands of years of human domestication, and he would especially eliminate intellectual training, because all the young needed to know was self-control, self-discipline and courage in the face of death. The young would redeem Germany from a failed past, he would make them hard, and the next time they would win.[19]

Freud once spoke of pity as a normal emotional barrier that brings the instinct to aggression or mastery to a halt at the sight of another person's pain. This is precisely what Hitler insisted had to be overcome, this unreflected willingness to stop, this failure to follow through ruthlessly to a logical conclusion, a failure that stemmed from the lack of something grand to believe in, something grand enough to evoke and sustain the ruthlessness. Hitler had watched good men ground up every day in that other holocaust, and for nothing. The Germans had lost because they were not tough enough, determined enough, merciless enough, they lacked a coherent communal ideology around which people could be rallied, and they had failed to identify and destroy the real enemy. And this last especially was where the traditional authorities were most culpable; the civilian leaders of the Second Reich had failed to protect the fatherland, they had allowed themselves to be penetrated and ruined, they had misperceived the nature of the struggle, the result above all of commitments to a false and deceitful civilized posture and morality. There were no further historical or moral obligations to that past, nothing left to revere, preserve, or restore.

Hitler was shocked and outraged by the defeat, but he recovered, got involved in politics, drew strength from his experiences, even the defeats, and relied finally on his own grandiose sense of mission to institutionalize the break with those representatives of inferiority and failure, and to repair the damage himself. But Hitler could and did act only in terms of his grandiose conception of racial struggle. It makes no sense, therefore, to argue that Hitler was interested in power for its own sake or that he was a nihilist destroying civilization for the pleasure it provided. On the contrary, Hitler was concerned with the fall from perfection, with human limitations, with freeing people imprisoned in a world of moral and bodily restraints. He wanted to break through to a new reality, the true one, he thought, and one more consistent with human potentiality and desire. Hitler had ideal aims and ambitions and he tried to organize the means to realize them. He was absolutely, murderously unscrupulous in the choice of means, but that is another matter.

It makes no sense either to argue that Hitler was a closet bourgeois, and that one day the mask and the brown shirt will fall away revealing gold teeth and a dollar sign. Hitler did not serve the interests of the bourgeoisie, he did not particularly admire them, and he was interested in their fate only insofar as they also served his racial aims. The bourgeoisie may still have believed in personality and hierarchy, but they did not know how to protect themselves. They had created a social order in which Jews could flourish and even dominate and had underwritten their own destruction; their day was over.[20]

Hitler realized that he could never get as many people as he needed to act exclusively on his racial anti-Semitism. He also wanted certain other problems solved, particularly the problem of economic independence from foreign sources of supply, preferably through conquest and the establishment of German hegemony on the continent. He therefore always offered his followers, including the bourgeoisie, a range of important secondary goals they could identify with and act on. At the same time, of course, the bourgeoisie still represented an interest in the society, they were still seeking to improve and strengthen their position, they were certainly grateful for the victories, and they never renounced the right to plunder. But Hitler was not one of them and he did not serve them.

It makes no sense, finally, to insist that all Germans understood their situation in the same terms, or that they really knew what Hitler was talking about.* Hitler meant to save the Aryan race, and his radical posture to the world was so thoroughly rationalized, integrated, and controlled, and from his own standpoint scientifically justified, that there was no need to keep his anti-Semitism public at all times, to react to every "provocation," or to rant before every audience. Hitler subordinated impulse to achieve a systematic conclusion, and he would talk to the German people about Christianity and the family just as coolly as he would treat with Poland, the Vatican, or the Soviet Union, just as coolly as he would trumpet his desires for peace or declare that Germany's aims in Europe were limited, or that he was a reasonable man and had too much respect for the national principle ever to violate it where others were concerned. Hitler was a skilled and disciplined politician. From time to time the discipline broke down and he would become strangely enraged. But he could easily rationalize such things: "Probably none of us is entirely normal," he said.[21]

* The genocidal wish was certainly not consciously shared throughout the population. Whether it was unconsciously shared is an impossible empirical question, although this has never stopped some psychoanalysts and psychohistorians from saying so. However, as such statements can neither be proved nor disproved, they have a negligible status and do not warrant further discussion.

Hitler's racist thoughts, his omnipotent world-saving and world-destroying themes and ideas, and his manner of confirming them, especially his absolute refusal to consider any information from any source that contradicted his racist views, may be described as paranoid. The clinical term is inelegant, but there is no mistaking the repeated identification of an international conspiracy mobilized by the Jews, implicating capitalism, Bolshevism, and Christianity, aimed at the destruction of the Aryan race. There is no mistaking either the manifest ability to assimilate all kinds of information to these thoughts, whether it was Britain's stubborn capacity to resist, or France's failure to resist, or revolt against Mussolini in Rome, or the "fatal weakness" of the Soviet Union. Hitler was quite convinced that the Aryan race, seeking to redress injustice and claim a rightful place in the world, was locked in a battle with destructive, annihilating Jews, responsible everywhere for the destruction of everything worthy and good, the outcome of which would decide the fate of mankind.[22]

However, Hitler's preoccupation with Jews was not necessarily what struck people with the greatest force at the time, and it was not what made Hitler a socially effective leader. Hitler's contemporaries were more struck by his grandiosity and his peculiar vulnerability when it was or appeared to be thwarted or denied. Hitler's thoughts and feelings about the Jews were explained away or seemed unworthy of serious attention because they ran counter to common sense. But the grandiosity and the vulnerability were puzzling and compelled attention in a way that Hitler's racist thoughts did not.

Hermann Rauschning very ably and perceptively commented on Hitler's grandiosity and his need to force the world into compliance with his wishes or suffer loneliness and emptiness as a result. Rauschning wrote that

> in the ecstasy of his speeches, or in his solitary walks in the mountains, he feels that he does possess this [power]. But in the many lethargic, empty hours he feels humiliated and weak. At such times he is irritated and unable to do or decide anything. He tries to acquire the semblance of creativity by endless talk. This requires an audience.

Hitler, according to Rauschning, needed constant reassurance and expressions of enthusiastic approval, depending on those around him to share his views. And he was so exclusively wrapped up in himself that he was incapable of genuine devotion. Rauschning observed Hitler's craving for sweets, as Hitler's secretary had also observed it during the war: Hitler stuffed himself with chocolates as Berlin burned, like Kafka's Hunger Artist, looking for the right food, hoping to fill himself up with good things and be free at last. Rauschning also observed the rage provoked by the thought that he might never experience that freedom ("He was an alarming sight").[23]

Alan Bullock had already indicated by the early 1950s the importance of Hitler's grandiosity (megalomania), his boundless egotism, arrogance, and vanity, his touchiness, his constant need of praise, his exalted sense of mission, his constant expressions of will, force, and ruthless determination, his inability to feel any remorse or regret for his savage acts, his sense that the gravity of his mission released him from ordinary constraints and exempted him from the civilized canons of human conduct. Bullock also referred to Hitler's interpretation of all events by the standard of how they affected him, a posture Bullock knew was confirmed by people who knew Hitler. Franz Halder had noted that

> even at the height of his power there was for him no Germany, there were no German troops for whom he felt responsible; for him there was—at first subconsciously but in his last years fully consciously—only one greatness, a greatness which dominated his life and to which his evil genius sacrificed everything—his own ego.[24]

Such observations occur too frequently and too consistently to be dismissed as inventions. Albert Krebs noted, for example, that

> the rejection from the army high command had not only hurt and angered Hitler, it had shaken his self-image, in its sensitive insecurity. To restore this self-image, it was not enough for him to refute the opposition objectively; no, he had to dream himself into the position of great men of power, the world movers of historical significance. Only thus could he become fully conscious of the difference between his own greatness and the nothingness of the

'dwarfs' opposing him. He needed this image in order to dull or shout down his own fear. Like so many tyrants, for Hitler, too, the words and deeds of his delusions of grandeur were defense mechanisms against the fear that he was, in the last analysis, unequal to his self-chosen tasks.[25]

Hitler's expressions of immense vanity, boastfulness, and inordinate strivings after grandiose ambitions on the one side, and his great sense of insecurity, his intense fears of weakness, passivity and humiliation on the other are characteristic of what psychoanalysts refer to as narcissism or a narcissistic style. It is this pattern of behavior that must be unfolded in order to encompass in one systematic frame of reference his thoughts about the Jews—which were peculiar even by the standards of some of the people around him—and his effectiveness as a leader.[26]

We can understand in these terms Hitler's ability to use "his" people as a means for realizing his grandiose ambitions and, given the ultimate failure, as a focus for his destructive rage. Hitler loved the German people for his own sake, not theirs, and their remarkable expressions of loyalty and devotion served mostly to confirm his own greatness. Hitler did not think about the Germans as they were, but as the embodiment of his racial ideals, and he became bitter and envious when reality did not conform to his expectations that he and his people were superior to all others.[27] Indeed, Hitler viewed Germany's failure as an attack on himself, expressing vicious recriminations over the weakness of the people or over the cowardice of the generals (characteristically envying Stalin's cleverness for having seen to this problem beforehand):[28] Their needs depleted *him,* their weakness was destroying *him.* When the German people failed to protect him (that is, from the pain of reality, which his closest advisers often did by withholding information or distorting it, or not contradicting him when they knew he was wrong), he became enraged toward them and ordered their destruction even as he was forced to consider his own.[29] An individual destined to lead the world, whose speech must be exclusively brilliant and whose every word was worthy of being recorded, who in fact had all but made himself the master of Europe, could hardly be expected merely to hand his sword over to the victors and suffer whatever fate they thought appropriate.

We can also understand in these terms the often-remarked and very important need of Hitler's to delay decisions inexplicably at critical moments, a vital aspect of Hitler's intuitive style of thought.[30] Hitler relied on his intuition, on the sudden insight, the inspirational flash, the interior revelation, which he never experienced without feeling himself superior and others (the objects of his wishfulness) as inferior and contemptible. Hitler then separated out the possibilities suggested to him by his intuition and pushed them to extremes. Hence, his constant resort to "either–or," and the corollary, "no capitulation": either a world worthy of his ambitions, or nothing. This mode of expression is ordinarily intended to mean that any conclusion will prove intolerable to one side or the other, that is, if both sides fear death equally.[31] But Hitler, who was forbidden to compromise with any vision less exalted than his own, feared failure more than death, and he was quite prepared to accept nothing.

Hitler posed the problem this way because he could not distinguish in his own mind what he was from what he did. Failure did not merely reflect on his decisions about inherently ambiguous situations, aspects of which were out of his, or anyone's, control. Failure reflected rather on his being, so that either–or meant that either he would win and be revealed as the greatest genius in history—or he would die, that is, he would sooner die than be revealed as an empty, helpless, wretched human being, sooner than suffer the shame and humiliation of defeat and the disintegration of the personal meaning of the world. Hitler delayed decisions, then, because his very being depended upon others acting in anticipated ways and if he failed to make the right decisions, failed to control reality, then his world would crumble. Every important decision implicated this vulnerability, so that victory or death were in fact the only alternatives.

Hitler's inability to think in relative terms, his inability to integrate good and bad, love and hate, in single individuals and groups, was the result of this constant tension between grandiose expectations and fears of weakness and failure. It was always this or that, never this and that. This pattern of splitting the world into unalterably opposed representations of good and evil stemmed from fears of emotional annihilation, the feared or anticipated death of his world. Hitler expressed grandiose and persecutory thoughts; he felt with great force that all good was in him and all evil external to him and he acted in the most extravagant ways on this feeling. But he never thought about guilt because it was never a matter of transgression. On the contrary, Hitler thought the world could only be grateful for what he had done.

Hitler claimed that to be effective one had to concentrate attention on a single enemy, encompassing all the evils of the world in that enemy so that people could be encouraged to act against a single force with a sense of prospective success and personal rectitude. Hitler also claimed that he had been terribly cheated and deprived and had learned to be hard and unsentimental toward his enemy. But Hitler's conspicuous lack of concern over the social and moral consequences of his actions, his virtual inability to think in such terms except in a mocking way, were not just the result of manipulative technique or hard knocks.

Rather, Hitler was afraid that he might not be able to realize his ideals, which meant to him the end of everything. This fear led to those vengeful assertions of his supreme and final power that we are familiar with. Clearly, Hitler was able to maintain his tolerance for events after 1942 by increasing the degree of his self-absorption and isolation. Hitler removed himself from the world more and more, narrowing his contacts while exacting the most grandiose vengeance on his sworn enemy. Hitler commented several times in 1942 on Jews laughing at the prospect of his defeat, that is, enjoying the spectacle of his disgrace. He obviously meant by these comments that if destruction was to be his fate it would be theirs, too. It must be emphasized, however, that anyone who thwarted Hitler's grandiosity was in danger. Hitler did order the destruction of Paris, as he also ordered the destruction of Germany.[32]

Hitler claimed that there were no restraints on the force of his will, and this was true in the sense that he was indifferent to consequences.[33] He would destroy the Jews, the Slavs, or anyone, in the attempt to force the world into compliance with his wishes. Hitler constantly invoked force of will, fanatical perseverance, as the source of liberating power, a reference actually to his own capacity to think exclusively in the service of his racial ideals. This in fact permitted a high level of integrated planning and activity, but in the particular sense that thinking was segregated from any moral deterrent. Hence, the ruthless movement in defiance of conventional logic and practice, the abandonment of all considerations of sympathy and feeling for those defined as the enemy.* Wishes ordinarily transformed and "socialized" were acted upon in this world of entitlement, and exalted, moreover, as representing the best in man restored to a rightful primacy.

Hitler's wishfulness, force of will, and capacity to act on racist ideals found a place in the disrupted world of Weimar Germany. His behavior was characterized by peculiar idiosyncratic features, but these nevertheless proved to be strengths for him and were recognized as such by others. Hitler's narcissism prohibited a withdrawal from the social world even though it was problematic for him, for example, because his grandiosity had to be confirmed continuously by an admiring audience. Hitler sought a life style that provided him with social importance and power over others. Hitler needed the audience, he preferred the spoken to the written word because he needed to see the effect, the audience mirroring his grandiosity through their exulting response. And conscious and assured of the effect, he would risk his life to make his conception of the world prevail.

In addition, one of the characteristic features of Hitler's narcissism was the peculiar vulnerability to selective reinforcement of native talents, which were then exploited all the more because they were highly regarded and rewarded by others and

---

* Hitler's ideals were ultimately threatened by the very vigor of his assault; he forced the opposition to stir itself. The tension between the sense of hope in the imminence of a grander future, in the creation of a world suitable to such highly regarded ideals, and the spirit of resistance that was finally engendered, which raised the prospect of defeat and annihilation, was resolved by the paradoxical mastery of technology in the service of death.

because they served also to mask shameful deficiencies in himself. Hitler was a speaker of considerable power, he was genuinely admired by others for this gift, he relied heavily upon it to great effect—and his deficiencies were correspondingly condoned or ignored.[34]

Hitler fought to be the center and creator of a world, and the power that animated him made him credible to others. This was particularly true in the crisis years when his omnipotence (in practical terms, his capacity to remain defiantly oppositional regardless of the odds, his apparently superb self-control, his confident determination, manifest particularly in a verbal capacity related to some ideal notion that he could talk people out of a sense of enforced passivity into renewed aggressive activity), complemented remarkably his public's fear of disruption and decline. Hitler talked about being abused, deprived, exploited, and frustrated, he pressed the idea that the Germans should be masters, not pariahs, he harped on fears of deprivation and hunger, on emptiness, on betrayal and abandonment in a world full of envious enemies, promising to turn passively experienced decline into active control, to recover and expand the effective power of the nation. Hitler promised that he would master reality and people believed that he could.

To be sure, archaic personal strivings were connected in his mind with the contemporary reality. But that enabled him to have a vision of Germans they no longer had of themselves. He always acted, too, as if the defeat of 1918 had never really occurred, and he quite sincerely told Germans that they had been betrayed, that their real inner core had not been affected. The sense of conviction with which he said these things appeared as strength that others might admire and respond to in a subordinate way.

It is important to understand, however, that Hitler became effective in this situation primarily because he addressed himself to real problems, no matter how "regressive" his solutions may have been or become, even to those who once supported him but subsequently came to see his solutions as mad and insane. Hitler's actions were oriented to reality, he promised solutions to real grievances and he was highly admired for that reason. Hitler became important because he was perceived as powerful and as part of a shared reality. And in describing Weimar Gemany as weak, powerless, defenseless, treasonable, and passive in the face of outrageous demands of alien forces, or in driving to make Germany great and independent as expressed in the unity of the people, in industrial and military might, and so forth, Hitler appealed directly to the achievement of commonly held, socially purposive goals as they were then conceived.

Hitler's apparently unswerving commitments and his uncompromising assertions of right and wrong, good and evil in the language of strength, will, activity, sacrifice, nobility, health, rootedness, community, retribution, and rebirth, provided masses of people with a sense of integration and purpose through submission to his authority. Such submission may have been at the cost of personal autonomy and insight, but this too was often an elevating experience as it defined membership in a valued community.

This emphasis on Hitler's narcissistic and paranoid traits is necessary because it is of more than biographical interest. These traits appeared and were expressed in a unique and idiosyncratic way, but they were systematically elaborated and became systematically useful in a particular social context. If Hitler's behavior can be conceived of as pathological, it was a pathology of ideals. And when the Weimar regime failed, it failed singularly to provide an environment in which the Germans could live up to and act on their ideals. This was the force that animated Hitler's efforts and it was the fear in others that he manipulated and resolved with the greatest effect. Political leadership may involve an individual's view of himself, but it is also inevitably a social phenomenon and the paths of this unique individual and the larger public crossed at this point.

Hitler thought he could become the greatest German in history, and that he had a world–historical mission to rescue Aryan blood from degeneration and corruption. He therefore constantly and with conviction held himself before the public as the embodiment of German ideals, he constantly posed as the representative of the German spirit and mission. This posture was necessary for his personal integrity and stability; but it complemented at a unique juncture in history the fears of masses of people, specifically in terms of their ability to live up to and act on their own conception of these ideals. Hitler became socially effective, his behavior and aspirations were systematically tied to the wider world, because he acted with force to defend what millions of people conceived of as most admirable, worthy, and German, and what they were convinced was grievously threatened.[35]

Hitler was under no illusion that he and the German public saw the world the same way. He assumed rather that the Germans would have to live under Nazism for two or three generations before it became the only world view that made sense to them. Hitler did not therefore trust the Germans to act on every goal he had in mind sooner or later to accomplish. He was determined to train an elite corps to see to that. The narcissism that elevated Hitler, fostered a belief in a peculiarly worthy mission, and enabled him to persuade people that his strength could rescue them from imminent failure also compelled him to try to realize this mission on his own terms; and his personal power was sufficient to recruit especially among those who could follow him in his step back from a world of symbols to a world of objects, and who would follow him wherever that led.

Hitler believed that modern struggle involved ideologies, not interests as conceived of in a narrow bourgeois sense. Interests of that sort could be compromised, ideologies never. Such a struggle, implicating the fate of civilization, the shape of the world to come, could be pursued only in the most pitiless manner. Hitler knew there were people who thought he was crazy because of his ferocious, single-minded intensity. But that did not matter because the winner of this struggle would make his vision of reality prevail, would impose his version of the truth. The winner would define reality and determine who was crazy. The loser would go to the wall. The only thing that mattered was success, and as his success promised the resurrection of man, any means were justified in its pursuit. In this way, will, resolve, dedication, discipline, sacrifice, and manliness were turned into the legitimation of conflict, aggression became an everlasting and sacred norm, and some Germans, at any rate, felt entitled to realize their racial and national claims, no matter what this might mean to anyone else.

Some Comments on Traditional Psychoanalytic Orientations to Social Action

1. Freud observed that people do not remember, they act out, recreating attenuated versions of threatening or damaging experiences in order to master them. Freud assumed that this kind of acting out of infantile and childhood conflict is systematically, inappropriately, and routinely repeated in current social situations. However, this assumption obscures the fact that social conflict may provoke psychic instability and require the kind of response Freud thought of as "regressive" in order to avoid even more dangerous kinds of breakdown. People strive to remain connected to the social world, the arena for the expression of interest, morality, and wishfulness, and such responses serve adaptive purposes in situations of conflict. They are perceived by the participants as appropriate and cannot be understood as merely neurotic or deviant. Unlike neurotic behavior, such responses will appear socially purposive and relatively free of conflict, no matter how they appear to outsiders.

2. Freud emphasized threatened or actual object loss as one realistic basis for both anxiety and depression, meaning that such experiences are inevitable in the early stages of development. He also observed that the feelings associated with the experience of loss are kept alive continuously by the fear of what the superior power of fate may bring. However, this must include realistic (as opposed to fantasied) threats of loss occurring in adult life, not only in a familial sense, but in a much wider symbolic and social sense. Moreover, the capacity to withstand realistic threats or objectively damaging events depends upon ideological and organizational ties that can be rendered dysfunctional or inapplicable to current needs. It is necessary for this reason to make a distinction

between the traditional psychoanalytic conception of "acting out" (a manifestation of unresolved intrapsychic conflict of whatever sort) and social action that serves as a means of sustaining a sense of continuity and personal integrity in a crisis situation, in terms of shared standards and expectations.

3. Freud's position fostered a sense of unity or uniformity of personality type because he would only consider one exclusive, drive-derived source of symptomatic expression. The phylogenetic logic that supported this simplistic scheme has long since collapsed; but the scheme retains its symbolic force among psychoanalysts and psychohistorians, in one form or another, and it does not seem to matter how differentiated and complex psychoanalytic knowledge becomes. It is ironical, moreover, that Freud actually understood the complexity of the problem, despite the fable of the totem murder and the confused and contradictory suggestion of a "psychical reality" independent of any social experience, either or both of which suggested a single content stemming from a single source.[36] In his report of the Dora case ("Fragments of an Analysis of a Case of Hysteria," 1905), Freud noted, for example, that a symptom can correspond to several meanings at the same time, can express several meanings in succession, can change its meaning or chief meaning over time, or the leading role can pass from one meaning to another. And as Steven Marcus has commented with respect to these observations, "The meaning in question may be a contradictory one; it may be constituted out of a contradictory unity of opposites, or out of a shifting and unstable set of them. Whatever may be the case, the 'reality' that is being both *constructed* and referred to is heterogeneous, multidimensional, and open-ended." [37]

The point is that all people live developmentally through experiences of separation, neglect, abandonment, seduction, disappointment, or of loving support, benevolent interest and concern, realistic constraint, and so forth. The aspirations and controls that are organized in terms of these experiences are constantly active, one or another may suddenly dominate attention, as a result, perhaps, of accidental personal experiences, or as a result of unanticipated, potentially disruptive social experiences.

The conflicts that derive from such experiences may or may not be resolved, they may or may not recur; such conflicts vary from one person to the next, they are not felt with the same degree of intensity, and they do not have a single source, content, or portent.

Hitler's behavior may be called paranoid, then, but not everyone who shares a paranoid conclusion in a stressful situation shares thereby the same paranoid disorder. Hitler's behavior may also be called authoritarian; but authoritarian behavior can be expressed in a variety of perseverative conceptualizations and also by otherwise "normal" people when social events interfere with the ability to act in familiar and culturally valued ways.[38] Hitler was himself an intuitive person, living and acting in anticipation of the fulfillment of racial visions, anxious to experience an ideal future; he was inattentive to administrative detail and indifferent to, and perhaps incapable of, methodical organizational work. He therefore could not have begun to realize his visions without planners, pluggers, and sluggers who acted out of a variety of motives.[39] Hitler personally responded to a disruptive situation with a sense of anger and injustice, mobilizing some people who responded similarly to the initial situation or to later ones, but mobilizing other people who responded out of a sense of loss, danger, or euphoric release, and still others who, impressed by the force of his convictions, decided not to wait things out, which they might otherwise have done, or who saw in the expression of power a favorable situation they might exploit to their own advantage.

4. The family has typically served as the sociological "middle term" between the kind of ontogenetically derived individual strivings Freud identified and systematic social action. The familial concept rescued psychoanalysis from Freud's failed logic—but it has also been understood for some time that the family is by no means a necessary and sufficient term and that the behaviors in question can stem from other sources. Thus, Else Frenkel-Brunswik wrote in 1954:

> Although in Germany a long history of authoritarian regimes is mirrored, and undoubtedly reinforced by, authoritarian family

and school structures, totalitarianism may well arise in countries with more permissive family atmospheres. Anxiety-inducing social and political situations such as economic depression and war can bring to the fore irrational elements and feelings of helplessness, and thus create susceptibility to totalitarianism regardless of how democratic the family situation might have been.[40]

The implication here is that the children raised in contemporary German families were systematically predisposed to totalitarian conclusions, while children raised in "democratic" families would have to be broken down by the force of specific situations. But in truth, we cannot infer the ability to engage in socially useful activity in the Nazi era or the willingness to participate in the Nazi movement, much less the willingness to participate in the Holocaust, from systematic child-rearing experiences in a particular familial context. The force of specific situations is determinative in either case. There is an implicit promise in all social organization, democratic or authoritarian, liberal or conservative, that one can at least approximate or strive to attain ideals in action at a level consistent with desire and capacity. The inability to remain effective in these terms as a result of disruptive social conditions must cause conflict between ego and reality regardless of any other dynamic conflicts that might also be present.

Society exists for people as a "background of safety," and it appears stable to them primarily in terms of the stability of ideological conceptions that explain and legitimate a variety of routine, everyday activities and occurrences, as well as more complex, abstract relationships. Abrupt, extensive, or prolonged environmental changes affect these activities and relationships, and hence the stability of ideological conceptions, producing sufficient tension so that anyone might respond in a rigid, inflexible, and ruthless way, no matter what the familial past was like. Any individual will experience distress when his learned confidence in social belonging, in his own and his group's virtues, is undermined, when shared conceptions no longer apply because familiar activities are threatened or are no longer available.[41] This is why it is necessary to understand the subjective effects of disrupted social conditions, why social movements are always heterogeneous in composition, and why charismatic leaders and their manipulations of ideological positions are of paramount importance. Leadership and ideology serve to bind socially and psychologically heterogeneous populations into relatively stable movements over time.

Notes

1. The incident is well described in John Toland, *Adolf Hitler* (New York, 1976), pp. 468–69.
2. Ernst Deuerlein, *Hitler: Eine politische Biographie* (Munich, 1969), p. 62. There have been chaotic situations in which charismatic leaders have not appeared, and whether such a leader turns up on the right, the center, or the left is an accidental factor, as all political possibilities are available and capable of being legitimated by such a leader's activity.
3. Hitler's proclamation of February 1, 1933, is reproduced in Henry Cord Meyer, ed., *The Long Generation: Germany from Empire to Ruin, 1913–1945* (New York, 1973), pp. 199–203.
4. The routinization of charisma actually refers in this sense to the stabilization of ideological expression identified by a particular language, implying then the internalization of a limited range of possible meanings that are sufficient to absorb the variety of interests, moral orientations, and fantasies in a population. The exclusiveness that characterizes ideological expression in the revolutionary period dominated by the great leaders stems from the primacy of moral as opposed to cognitive styles in the manner of expressing ideological convictions and with respect to the question of how it is that people must live.
5. On the anonymous–famous paradox, see n. 2 in the introduction; on the wise child paradox, *Mein Kampf*, trans. Ralph Manheim (London, 1969), pp. 5–15.
6. On Hitler's psychological turmoil, *Mein Kampf*, pp. 47–60.
7. On the evidence pertaining to Hitler's developed racist anti-Semitism see Rudolph Binion, *Hitler among the Germans* (New York, 1976), pp. 2–3.
8. *Mein Kampf*, p. 53.
9. *Hitler's Secret Conversations*, trans. R. H. Stevens and Norman Cameron (New York, 1953), pp. 115, 126–27. Karl Saller, *Die Rassenlehre des Nationalsozialismus in Wissenschaft und Propaganda* (Darmstadt, 1961), p. 129.
10. Binion, *Hitler among the Germans*, p. 2.
11. Quoted in Werner Maser, *Hitler: Legend, Myth and Reality* (New York, 1971), pp. 115–116.

12. Erich Ebermayer, *Denn heute gehört uns Deutschland . . . Persönliches und politisches Tagebuch* (Hamburg, 1959), pp. 24, 49, 76–77; Joel König, *Den Netzen entronnen: Die Aufzeichnungen des Joel König* (Göttingen, 1967), p. 56.

13. Maser, *Hitler*, p. 116.

14. The problem for historians is that though many other people went along with Hitler's decisions, based on these conclusions, we can never tell for what reason, and this is true even for the camps and not to the exclusion of opportunism. People may have followed as a result of distorted visions that resembled, or only complemented, Hitler's, but the possibility for knowledge ends when we try to trace this back through the different levels of personnel involved. This is why "homogeneity of unconscious motivation" was such an attractive prospect. The "id mythology" closes the gap between available evidence and the needed knowledge in a "scientific" language that seems to have an empirical reference. This accounts for the "heads I win, tails you lose" posture of some psychoanalytically informed writing. If someone produces documentation tending to confirm homogeneity of motivation, that is good. If someone else produces contrary documentation, that does not matter because the problem is unconscious anyway. On the distinction between viewing people in terms of what they do or what they are, see Edith Jacobson, *The Self and the Object World* (New York, 1964), p. 146.

15. See Binion, *Hitler among the Germans*, pp. 14–33, for a very concentrated dose of the concrete body language; and Maser, *Hitler*, pp. 116, 163, 170–71, for more.

16. *Mein Kampf*, pp. 21, 114, 142.

17. *Hitler's Secret Conversations*, pp. 115, 384. George Mosse, *The Crisis of German Ideology* (New York, 1964), p. 302. Hitler was convinced that this discovery of the "Jewish virus" was one of the great discoveries of world history. People lived in a natural state before the Jews disrupted it and would return to that order when the Jews were gone. *Hitler's Secret Conversations*, pp. 255, 269. Dietrich Bronder, *Bevor Hitler kam* (Hannover, 1964), pp. 334–337.

18. Maser, *Hitler*, p. 116.

19. Hermann Rauschning, *Hitler Speaks* (London, 1939), pp. 247, 241.

20. On Hitler's feelings about the bourgeoisie, see *ibid.*, pp. 30, 48; *Mein Kampf*, pp. 138–39, 159, 198, 225, 309; *Hitler's Secret Conversations*, pp. 49, 87, 202. Hitler said in passing once that Himmler would one day be the greatest industrialist in Germany. *Ibid.*, p. 105. In any case, Hitler made snide comments about the bourgeoisie all the time. The notion of Hitler as a representative of the bourgeoisie stems from an orientation in which there are only two kinds of people in the world, workers and bosses, and if Hitler was not obviously one, he must somehow have been the other, whether it is obvious or not.

21. *Hitler's Secret Conversations,* p. 127.

22. Scholars, and sociologists particularly, have always been skeptical of references to the leader of a great movement or to the beliefs or style of a movement as "paranoid," because it implies that behavior is symptomatic, that is, repetitive and stereotyped but also idiosyncratic and personal. And not only is it unclear how such personal resolutions to conflict become relevant to others, but these symptomatic resolutions are also typically rather rudimentary compared to the elaborate, often skillfully executed plans and manipulations of such a leader. The level of organizational skill, integration, and synthesis, the very insight into situations so often displayed, seem to belie the tag "neurotic" or "deviant." The plans and actions of the dictator and his group are too systematic, all-encompassing and "rational" in their context to be labeled paranoid.

The problem with the paranoid concept for sociology lies in Freud's initial explanation of the fantasy (sexual) origins of such dynamics; the problem, however, clearly needs to be examined further. To begin with, paranoid behavior may be viewed as a means of protecting organized activity, as a way of avoiding withdrawal from the social world, characterized by an affective style (punitive rage based on rigid polarization of the world) and a cognitive style (hostile suspiciousness of motives and a desire to penetrate superficial appearances in order to locate the true, underlying reality). Further, paranoid behavior is derived from or related to (in the sense of clinical progression or as a substitutive phenomenon) a variety of different conditions. Identifying paranoid traits as such is therefore not as crucial as identifying the total constellation within which they may appear. Paranoid dynamics frequently occur, for example, in relation to such other tendencies as narcissism and depression, both of which are peculiarly related to socially disruptive situations, either in a developmental sense, or as a direct result of such situations (which facilitate the expression of ideal images and ambitions for a narcissistic individual and hence the reorganization of socially purposive behavior for a depressed individual). In short, as the literature since Freud amply discloses, paranoid ideas may be socially derived, they may be used to develop a socially integrated world view, facilitating real interactions with people. At the same time, not everyone who shares a paranoid conclusion in a stressful situation shares thereby the same paranoid disorder, as I have already stated.

I should point out that Frederick L. Schuman, *The Nazi Dictatorship* (New York, 1935), pp. 95–130, had already identified the Nazi phenomenon as paranoid in style and the sociologist Theodore Abel, *Why Hitler Came to Power* (New York, 1938), pp. 187–194, criticized Schuman on the grounds previously noted. Alex Inkeles, "The Totalitarian Mystique: Some Impressions of the Dynamics of Totalitarian Society," in Carl J. Friedrich, ed., *Totalitarianism* (New York, 1954), p. 93, returned to this problem; and Norman Cohn, *Warrant for Genocide* (London, 1967), pp. 258–267, referred to Nazism as paranoid again. This problem must be resolved—and can be—as the infomation will not go away. In March 1944, a memorandum was submitted to Goebbels' ministry, the subject of which was the fear that people might no longer consider the Jews as an immediate threat or that the Jewish question was settled. The report indicated that, leaving six million Jews aside, there were still ten or twelve million left in the world and the German public must be mobilized against them, as they were still responsible for mobilizing opposition in Britain, America, and the Soviet Union. The Jews were the link tying these societies together in struggle. The author feared especially for young officers for whom the Jewish question might just become mere history. Léon Poliakov and Josef Wulf, *Das Dritte Reich und seine Denker*, (Berlin, 1959), pp. 461–468. Also see the statements published in 1944, on ideological education with respect to the role and aims of the Jews in Walther Hofer, *Der Nationalsozialismus: Dokumente 1933–1945* (Frankfurt a/M., 1957), pp 34–35; and Josef Ackermann, *Heinrich Himmler als Ideologe*, (Göttingen, 1970), pp. 155–160. Further examples: "Only if those with the slightest trace of Jewish blood are considered as Jews can we hold the line. Even if we are occasionally moved to make an exception in order to alleviate individual cases of hardship, there can be no budging from the principle lest our sentimentality should erode the . . . ideals of the Führer. The systematic extirpation of all Jewish blood that has contaminated the nation is the only salvation, the only positive eugenic step toward racial recovery." Dr. J. Hartmann, Leipzig gynecologist, quoted in Jacob Lorsch, "The Nazi Misuse of Mendel," *The Wiener Library Bulletin*, vol. 23, no. 1, New Series, No. 14 (Winter, 1968–1969):31. Or Gerda to Martin Bormann; September 8, 1944: Every child must realize that the Jew "is the absolute evil in this world." Wherever there is an Aryan who wants to work hard and cleanly and live according to the laws of his race, the Eternal Jew will try to prevent it and to annihilate all positive life. H. R. Trevor-Roper, *The Bormann Letters* (London, 1954), pp. 105–106.

It should be clear from the outset that references to a paranoid style do not imply a psychotic or near-psychotic condition. More specifically, maintaining a sense of stability in stressful situations in these terms is not exclusively characteristic of "sick" people; rather, normal people will behave similarly as we may observe in fleeting moods of anger, sadness, etc. On the relationship of masochistic, sadistic, and/or obsessive–compulsive behavior to paranoid dynamics, see Jules Nydes, "The Paranoid-Masochistic Character," *Psychoanalytic Review*, vol. 50 (1963): 215–251; and David Shapiro, *Neurotic Styles*, (New York, 1965), pp. 54–107. The relationship of depression to paranoid dynamics is discussed in David W. Swanson, Philip J. Bohnert, Jackson A. Smith, *The Paranoid* (Boston, 1970), pp. 286–290; Edith Jacobson, *Depression* (New York, 1971), pp. 92–95; and Elizabeth R. Zetzel, "Depression and the Incapacity to Bear It," in Max Schur, ed.,

Drives, Affects, Behavior, 2 vols. (New York, 1965), 2:243–274. The relationship of narcissism to paranoid dynamics, not discussed to my knowledge in the Swanson, Bohnert, and Smith text, is nevertheless discussed by Annie Reich, "Pathologic Forms of Self-Esteem Regulation," *The Psychoanalytic Study of the Child,* vol. 15 (1960):230–231; and by Otto Kernberg, "Factors in the Treatment of Narcissistic Personalities," *Journal of the American Psychoanalytic Association,* vol. 18, no. 1 (January 1970) : 56–58; and John M. Murray, "Narcissism and the Ego Ideal," *ibid.,* vol. 12 (July 1964) :495–510. On the paranoid response in borderline pathology see Melvin Singer, "The Experience of Emptiness in Narcissistic and Borderline States," Part II, *International Review of Psychoanalysis,* vol. 4, no. 4 (1977) : 476. Specific empirical studies of the relationship of paranoid dynamics to politics are rare and tend to be heavily ideological. See, for example, Henry A. Alker, "A Quasi-Paranoid Feature of Students' Extreme Attitudes against Colonialism," *Behavioral Science,* vol. 16, no. 3 (May 1971): 218–221. For a review of the experimental evidence on paranoid dynamics see Howard M. Wolowitz, "The Validity of the Psychoanalytic Theory of Paranoid Dynamics," *Psychiatry,* vol. 24 (November 1971): 358–371. Of particular interest on this subject are Leon J. Salzman, "The Paranoid State: Theory and Therapy," *Archives of General Psychiatry,* vol. 2, no. 6 (1960) : 679–693; D. A. Schwartz, "A Re-View of the 'Paranoid' Concept," *ibid.,* vol. 5, no. 1 (1963):15–21; Kenneth L. Artiss and Dexter M. Bullard, "Paranoid Thinking in Everyday Life," *ibid.,* vol. 14 (1966):89–93; Norman Cameron, "The Paranoid Pseudo-Community," *American Journal of Sociology,* vol. 49 (July 1943) : 32–38; and Cameron, "The Paranoid Pseudo-Community Revisited," *ibid.,* vol. 65 (1959) :52–58; Edwin M. Lemert, "Paranoia and the Dynamics of Exclusion," *Sociometry,* vol. 25 (1962) : 2–20.

23. Rauschning, *Hitler Speaks,* pp. 89, 219, 258–59; 256–57, 260. Hitler could in fact express genuine concern for people—if they loved him as he loved himself, or if they were killed or injured in his service. In any event, no one behaves one way all the time. Hitler's inordinate craving for sweets and the specific statement of his secretary is reported in Robert G. L. Waite, *The Psychopathic God* (New York, 1977), p. 413. Hitler's rage has of course often been commented on. Hitler did manipulate his outbursts, he did use the rage to cow people or to close conversations he no longer wanted to pursue. But Hitler was also often out of control, especially when his narcissism was affected by suggestions of failure or weakness, by contradictory information that upset some wish, or by the observation that people and things did not serve exclusively to enhance his own grandeur. The eminent surgeon Ferdinand Sauerbruch described Hitler's rage when he got Hitler's dog to play with him and to be friendly. According to Sauerbruch, Hitler shrieked with pain, demanded to know what Sauerbruch had done to the dog, insisted that the dog was no good to him anymore, and reminded the surgeon, who had flown to the Russian front to examine Hitler, that he,

Hitler, could have him killed. Sauerbruch claims that he got out of this awkward situation by telling Hitler that he simply understood animals with the insight that Hitler understood people. Again, this kind of observation is reported too often to be dismissed as invention or convention, and it is too consistent with the reactions of narcissistic individuals just to be coincidental. Ferdinand Sauerbruch, *A Surgeon's Life,* trans. F. G. Renner and Anne Cliff (London, 1953), p. 244. It should be noted that Hitler could tolerate real information and he could suffer contradiction; he took Speer's defiance, after all. But this depended upon who he was talking to, what the information concerned, and how he felt at any point. If Hitler had not also managed a high degree of control he could not have succeeded to the extent that he did, nor could he have gotten others to go along, of such quality, in such numbers, over such a long period of time.

24. Alan Bullock, *Hitler: A Study in Tyranny* (London, 1959), pp. 520–21, 686 (grandiosity); 336, 361 (egotism); 350 (touchiness); 614–16, 732 (the sense of mission); 349–350 (ruthlessness); 348, 352, 434, 724 (the absence of remorse, of scruples); 350, 632 (interpreting events by the one standard of how they affected him); and Halder's comment, p. 707. Bullock also noted Hitler's "uncanny intuition," the gambler's instinct, the either–or formulation, and the other side of the coin, the hesitation at critical moments, pp. 341, 343, 521, 346, 353, 618. Bullock took Rauschning's observations seriously (p. 346, n.) but then also followed him in concluding that Hitler really represented the revolution of nihilism. The racial doctrine was compelling for Hitler, and for others as well, even if it was cognitively distorted and even if the anticipated results could never have been achieved. The fact that the whole thing was self-defeating was not apparent, say, in May 1940, and the fact that it was wishful does not make it any the less real, as the fact that it violates common sense does not make it any the less appealing. Rauschning tended to deflect attention from Hitler's racism, which he thought was nonsense, he was angry over Hitler's betrayal of what he considered the potentially virtuous core of National Socialism, and he sometimes tried to liven up the narrative with allusions to Hitler's "morbid lusts," which he had no first hand information on and could not be specific about. But as far as Hitler's grandiosity was concerned he saw that clearly enough. He just has to be read with care.

25. Albert Krebs, *The Infancy of Nazism: The Memoirs of ex-Gauleiter Albert Krebs* (New York, 1976), p. 176. This posture of Hitler's is so well known that it has become itself a matter of common sense observation. Ralph Manheim noted in his translation of *Mein Kampf* that the logic of the text is purely psychological and that "Hitler is fighting his persecutors, magnifying his person, creating a dream world in which he can be an important figure." "Translator's Note," *Mein Kampf,* p. vii.

26. Erich Fromm discussed Hitler's (and Stalin's) narcissism in *The Heart of Man* (New York, 1964), pp. 66–76; and *The Anatomy of Destruc-*

tiveness (New York, 1973), pp. 406–07, 414–15. However, once Fromm located narcissism as Hitler's animating force, as opposed to the sadomasochistic character structure discussed in *Escape from Freedom,* he could no longer make sociological sense of the situation. In strictly clinical terms, John M. Murray characterized narcissistic behavior in terms of a "belief in omnipotence, a vital need for a feeling of magical control of a situation, cold brutality when this feeling of power is challenged, the narcissistic right to any kind of act which will restore power and entitlement, lack of pity and utter disregard for the effect of narcissistic acts on others and the classical reactive response to critical situations by the all-or-none law. Dominance and control must be immediate and absolute for the ensuing anxiety threatens him with complete destruction, and he reacts immediately to restore his omnipotent world at any cost to others." This clinical statement might well have been abstracted from a biography of Hitler; it could hardly have been more explicit or correct. John M. Murray, "Narcissism and the Ego Ideal," *Journal of the American Psychoanalytic Association,* vol. 12 (1964): 504–505.

27. As Speer commented, Hitler decided that he could leave Paris as it was because a grander, more glorious Berlin would overshadow it. Berlin was to become a capital on a world-historical scale, a match for ancient Rome and Babylon, a reflection of imperial grandeur, comprised of the largest, the tallest, the most resplendent buildings, the envy of the world, cowing visitors who would gasp with awe, etc., etc. But when Berlin was being bombed into rubble, Hitler ordered Paris detroyed. If it was not his and his was not the envy of the world, then there was nothing left but to destroy it. See Fromm, *The Anatomy of Destructiveness,* pp. 396–397; Fromm cites other relevant literature, in addition to Speer's observation.

28. Stalin was a beast, but a magnificent beast, proof that even subhumans could be organized for conquest if the proper leadership was available. *Hitler's Secret Conversations,* pp. 534, 541, 476. Ackermann, *Heinrich Himmler als Ideologe,* p. 206.

29. Percy Ernst Schramm, *Hitler: The Man and the Military Leader,* trans. Donald S. Detwiler (London, 1971), p. 28, a reference to Hitler's comment, 27 January 1942: "If the German people are not prepared to give everything for the sake of their self-preservation, very well. Then let them disappear." *Hitler's Secret Conversations,* p. 210. On the subsequent order to destroy Germany, Hofer, *Der Nationalsozialismus: Dokumente,* pp. 260–261.

30. The relevant instances are cited in Joseph Nyomarkay, *Charisma and Factionalism in the Nazi Party* (Minneapolis, Minn., 1967), p. 42; see also the reference to Alan Bullock, n. 24 in this chapter.

31. Anton O. Kris, "Either–or Dilemmas," *The Psychoanalytic Study of the Child,* vol. 32 (1977): 91–117. This, as Kris comments, is why either–or becomes neither–nor in practice.

32. Hans Frank said that Hitler's hatred knew no bounds, and this was true for individuals or groups or "whole races and nations." *Im Angesicht des Galgens* (Munics, 1953), p. 363. This included eventually the SS and even an individual like Sepp Dietrich, whom Hitler had at one point described as "a phenomenon," "a national institution," one of "my oldest comrades in the struggle." It did not help—Dietrich had failed somehow and as Hitler could not encompass contrary tendencies in single individuals or groups, Dietrich fell out of favor. See Bullock, *Hitler: A Study in Tyranny*, p. 711.

33. Rauschning, *Hitler Speaks*, p. 37. Karl Heyer, *Der Staat als Werkzeug des Bösen* (Stuttgart, 1965), pp. 47–50.

34. Regression is never total, not even among psychotics. It does not imply that one is disoriented, disorganized, or unable to cope with reality. There is a selective process of observation and construction; some, not all, ego functions are impaired. And insofar as one may thereby express strength, demonstrate an oppositional capacity, appear morally incorruptible, especially in a time of crisis, one can be a political leader. There is no contradiction between this behavior and the ability to run a complex industrial–military system. On the contrary, because industrial and military capacity are criteria of strength and vital to self-image and to the pursuit of ideals, the opposite is true.

35. The expression of narcissism may be described in Hitler's case as an idiosyncratic instance, but it became important to many other people who experienced distress in this area because of socially imposed, situationally specific conditions. Hitler's personal resolutions therefore became sociologically relevant. That is, socially disruptive conditions provided a suitable environment for the expression of narcissistic conflicts, which are characterized by exalted self-image and exaggerated commitment to cultural ideals and by the wish to be admired in these terms. This self-image can only be maintained, however, by forcing others into compliance with such wishes and the narcissistic individual will struggle with the greatest intensity to make himself socially effective. The convinced, fearless, and inflexible expression of ideals makes such an individual believable to others if they are threatened in their own ability to live up to or to act on ideals, which was precisely the situation in Germany. It is possible to make a broader statement on charismatic leadership in these terms, but I will not attempt that here.

36. See Fred Weinstein, "On the Social Function of Intellectuals," in Mel Albin, *et al.*, eds., *New Directions in Psychohistory* (Lexington, Mass., 1980), pp 3–19.

37. Steven Marcus, "Freud and Dora: Story, History and Case History," in Theodore Shapiro, ed., *Psychoanalysis and Contemporary Science*, vol. 5 (1976):83. Italics added.

38. Rigid thinking and splitting the world into good and bad as ex-

clusive polarities is a characteristic of manic behavior, as is an incapacity for emotional give and take. Manic behavior is characterized by a denial of weakness and dependence as well as by noisy demonstrations of courage, provocative rivalry, and illusions of omnipotence; moreover, anger may be expressed through the repudiation of restraints normally imposed by family and society. Projection also appears in the manic defense against depression; such an individual may feel himself loved and admired by everyone or, in paranoid fashion, feel mistreated and therefore entitled to do whatever he likes without any regard for others. And this behavior may be rationalized as fulfilling some ideal purpose. The paranoid is greatly concerned with power, the manic depressive less so. But there is a connection possible between the two through the expression of such themes as rebirth. The former rescues the latter who is "reborn" with hectic convictions of strength and high purpose. In any event, there is no need to think in terms of a single, or even a dominant, personality type, in contradiction of the evidence. Meyer Mendelson, *Psychoanalytic Concepts of Depression* (Springfield, Ill., 1960), pp. 45–76. David Gutman, "The Subjective Politics of Power: the Dilemma of Post-Superego Man," *Social Research*, vol. 40, no. 4 (1973): 570–616, especially p. 600. The literature on paranoid dynamics is quite extensive and there is a text available, Swanson, Bohnert, and Smith, *The Paranoid*. On paranoid dynamics, see also n. 22 in this chapter.

39. That is, we may think in terms of a cognitive reality of process and ideas, a feeling reality of emotions and memories, a sensation reality of concreteness and immediacy, and an intuitive reality of visions and ideal ambitions. Hitler was intuitive in style, but he could not have succeeded without the other capacities being available to him. Harriet Mann *et al.*, "The Psychotypology of Time," in Henry Yaker *et al.*, *The Future of Time* (London, 1971), pp. 170–172.

40. Else Frenkel-Brunswick, "Environmental Controls and the Impoverishment of Thought," in Friedrich, ed., *Totalitarianism*, p. 177.

41. There is evidence to indicate that unusually stressful events will produce symptoms of distress in a large proportion of the population exposed to them. Such symptoms should be seen as elicited by events and should not be seen in a systematic or modal sense as indicative of a single underlying psychic reality stemming, for example, from one fundamental family experience. Barbara S. Dohrenwend, "Social Class and Stressful Events," in E. H. Hare and J. K. Wing, eds., *Psychiatric Epidemiology: Proceedings of the International Symposium Held at Aberdeen University, 22–25 July 1969* (New York, 1970), pp. 22–25; G. W. Brown and J. L. T. Birley, "Social Precipitants of Severe Psychiatric Disorders," *ibid.*, pp. 321–325. Barbara S. Dohrenwend and Bruce S. Dohrenwend, *Social Status and Psychological Disorder: A Causal Inquiry* (New York, 1969). See also Swanson, Bohnert, and Smith, *The Paranoid*, p. 278.

4

Nazism as an Ideological Movement: Conceiving the Final Solution

Hitler claimed that the victory of Nazism was the victory of a racial world view and that Nazism would radically affect the way people lived. But not every Nazi was equally or even especially concerned with racism, and there was more than one orientation among those who were. Nazism in practice was anything but a monolithic enterprise evoking loyalty to a single standard, a situation reflected in the organization of the Nazi state, which was replete with competing and conflicting interests, jurisdictions, and lines of authority.[1]

This situation has led to a certain confusion concerning the role of Hitler's racial world view in the Nazi state. It hardly seems credible even yet that Hitler's brand of radical racism could have been the primary purpose for which Nazism was mobilized. Activity oriented to class, national, or self-interest always appears "real" in a way that the rationalizations that justify them do not. The rationalizations of a shocking and bizarre racism seem even more remote from any real interest, and the Nazi state is therefore more often explained as a bourgeois phenomenon, or as the expression of German militarism organized for the sake of conquest, or even as power mobilized for its own sake.[2]

However, it was not possible for Hitler to have mobilized himself and others for such activity in the absence of a synthesized, coherent world view that explained and justified its purpose. All activity is integrated and controlled by a network of rationalizations at the individual level by what Freud referred to as defense mechanisms and at the social level by what he referred to as internalized morality, or, equally valid, by what Marx referred to as ideology.[3] But while the "choice" of defense mechanisms (like the "choice" of neurosis) is an idiosyncratic phenomenon, internalized morality or ideology is systematic and social, controlling the latitude of legitimate responses in terms of shared standards and expectations.

Both Freud's and Marx's concepts imply that unconscious thinking undergirds systematic social activity, although Freud saw in the most decisive way that such thinking is inevitable and necessary because the relationship between individuals and social order is too demanding and problematic.[4] That is, people cannot assimilate in consciousness all the events that are constantly evoking memories of loss, separation, deprivation, and neglect, or of reparative relationships that are no longer appropriate or available. The ability to define and redefine oneself as worthy, admirable, and good, or one's behavior as adequate and continuous, can be sustained only if the significance of events is to some extent repressed, and if the internalized morality or ideology remains appropriate to an explanation of what is consciously understood. "Ego" is in a constant state of "deformation" as a routine, normal occurrence, and that is why, by way of reference to unique, idiosyncratic behavior, we have a psychopathology of everyday life. But "ego" is also in a constant state of synthesis, and when the ideological basis for the synthesis is challenged by adverse or conflictual events, we also have systematic and ruthless social behavior.

The capacity for ideological synthesis is indispensable for the ability to act, as people are enabled thereby to absorb and rationalize discontinuities of experience. Of course, any individual may randomly prove unable to do that for idiosyncratic reasons. But the inability of groups to do that can stem only from environmental failure that renders the ideological synthesis inapplicable to an explanation of events, leading to the kind of painful experience described earlier. Any real event that results in a discrepancy between valued aims and ambitions and the social means to realize them, or between the most modest personal standards and real achievement, fosters such experience and compels a search for ideological legitimations.[5]

Recent history has amply demonstrated that people will go to the most extraordinary lengths to sustain or impose an ideological orientation that legitimates valued activity. And if they cannot justify this in terms of objective or "true" connections, they will not hesitate to fabricate false ones. The "stab-in-the-back" legend in Germany is a pertinent and prominent example. But it has often happened in history that a class or group striving to retain or to seize power retrospectively falsified their past and distorted their motives in order to justify some current action. The reason that such false connections appear integrated and legitimate is that they serve a threatened sense of adequacy and continuity and so are not affected by realistic objections.

There is no possibility of organized social activity without an internalized morality or ideology to legitimate it. Indeed, ideology is the transcendant phenomenon of everyday life,[6] not as the causal or motivating factor in social activity, but as the link between organized mental activity, including unconscious activity, and the organized social world. Ideology is the structured and systematic form of high and low level (theoretical, philosophical, and common sense) language that people employ without reflection to explain all kinds of anticipated and unanticipated, favorable and unfavorable events.[7] Ideology allows society to absorb a variety of idiosyncratic strivings through conventionalized language, the meaning of which is assumed to be commonly shared. Ideology serves to legitimate realistic social activity, focus heterogeneous cognitive, moral, and wishful ambitions on the completion of social tasks, and define the range of appropriate behaviors in terms of those tasks.[8] Ideology, not class, is the most important variable in the explanation of social stability and social change.[9]

The ability of working class and Catholic parties to hold out against the Nazis in the face of objectively damaging events underscores the importance of this ideological process in a particularly interesting way. Both groups resisted the pressure of events relatively well because they continued to believe in the moral legitimacy of their respective orientations and because these orientations remained adequate to an explanation of events. Moreover, not only did the Catholic (Center and Bavarian People's) parties cut across class lines, but the Nazis found support among Protestants in the same strata that remained loyal to the Catholic groups.

The question is not, therefore, whether Nazism was an ideological movement, but how Nazi ideology could have absorbed the interests, moral commitments, and wishful expectations of a heterogeneous population, linking a variety of aims, ambitions, and potentialities, lending the illusion that most of the population, and particularly traditional conservatives and racial radicals, were talking about the same thing most of the time. Nazi ideology was sufficient for this purpose because of its particular content and because the content was sufficiently ambiguous to allow for a latitude of interpretation, that is, to allow people to supply whatever meaning was important for them.

The content of Nazi ideology included four basic themes, which may be conceived of as follows: The Nazis were past oriented, they could not envision a future without significant aspects of the past restored or elevated to a primary cultural position, meaning that cherished psychological positions would not have to be surrendered; the Nazis exalted struggle as a first, unalterable principle of life, a fundamental and irreversible mode of treating with the world, meaning that enforced passive submission to events could be avoided, and that aggressive activity was desired and legitimate; the Nazis cathected and exalted experiential and affective functions of character, as opposed to cognitive functions, dwelling ceaselessly on instinct, will, intuition, and "the blood," meaning that emotional responses were superior to rationally disciplined responses in the political sphere of activity; the Nazis valued and sought after a hierarchical social order on principle, they were or meant to be rigidly exclusive with regard to the lines of authority among themselves, and they were utterly contemptuous of alien people, meaning that Germans were peculiarly deserving and worthy, while others were not.*

There was, then, an easily identifiable, internally coherent Nazi ideology that served to restore a sense of adequacy and continuity to a large number of people in a crisis period, separating them from the immediate impact of events, orienting them away from exclusive preoccupation with present personal and cultural failure. The Nazi ideological appeal, linked by language to a traditional German conservative ideology, directed attention to the virtues of the past and, by promising to recover and restore those virtues to a rightful primacy, encouraged people to believe again in a future. Moreover, the implied latitude of interpretation rendered this ideology effective over time, as it could have appealed to interest, morality, or wishfulness, it could have been interpreted in terms of what people do as opposed to what they are, or in an abstract as opposed to a literal, concrete way.

Hitler could not have been all things to all men, nor did he want to be, since it was necessary at least to establish the opposed ideal against which people could define themselves with some

---

* In order to underscore the implications of the content of Nazi ideology, I will briefly contrast it with what I consider the equivalent effective content of Bolshevik ideology: The Bolsheviks were future-oriented, and aside from their own revolutionary tradition there was no part of the past to which they were committed in a conscious and determined way: the Bolsheviks upheld the virtues of harmonious, nonconflictual social activities, always interpreting their own aggression as reactive to the transgressions of a hostile world, viewing aggression as an unavoidable behavior stemming from a particular kind of social order. Aggression was not intrinsic to them, nor was it everlasting; the Bolsheviks cathected cognitive orientations, standing in their own minds for disciplined, reasoned, objective commitments to action. They were loyal to their bourgeois origins in this regard, they did not trust emotional responses and did not initially seek to evoke them; the Bolsheviks valued and sought after an inclusive, nonhierarchical social order on principle, and they were convinced that the distinctions the Nazis referred to were a function of a particular form of social organization and would disappear when that form was superseded. Needless to say, having confused a moral orientation with predictive theory, the Bolsheviks lost control over events and they had to sacrifice these ideological ambitions. The reasons for failure are historical and social and the response to failure may be understood in the terms employed here.

degree of unity. But Hitler did nevertheless have to be many things to many men, and he was easily capable of changing his style or his message to suit different audiences, manipulating the various criteria cited above. The trick was to get people to believe that he would serve their interests primarily or work to realize their ambitions before all others. And his advantage was, given the power he expressed in his person and that he was also able to mobilize in the brown battalions, people thought he might.[10] It remains now to see how this ideological process unfolded in practice.

Orientations to Past and Future

Karl Mannheim pointed out some time ago that conservatives tend to be past oriented.[11] In his view, traditional feelings and longing for traditional modes of behavior (hierarchy, dependence, exclusion, political passivity in the decision-making process) were retained within the rationalizing bourgeois societies by those social and intellectual strata that remained outside of or only grudgingly accepted the capitalist organization of society. The traditional aspirations were kept alive among the nobility, the peasants, and the petty bourgeoisie, those groups most likely to regret the passing of old ways and loyalties. Such groups tended in any case to reject the possibility for increased personal spheres of activity offered by the bourgeois societies, though many or even most of the people involved were oppressively exploited by their own standards in the traditional societies.

The problem was that the centralizing nation–state, ultimately dominated by the bourgeoisie, required a more complex and abstract kind of productive and distributive network, and managed to impose its power only after defeating recalcitrant classes (peasants, artisans) and localities (cities, provinces). For the defeated who could not stop the economic and political processes that so drastically affected their lives, the only alternative was, finally, loyalty to the wider, national society. Thus, the social and institutional background, which was by now in fact more remote and abstract, became all-important, and was often enough reinterpreted in terms of an idealized version of the older local background. This process, and the additional fact that religion, especially in its Protestant forms, could not adequately explain or justify the new social relationships or the psychological consequences of change, constitute the psychosocial origins of modern nationalism. The nation–state, however that entity was defined by such individuals and groups, became more important for them in time than the older localities had been. The loss of this national symbol would therefore have been felt more intensely because people actually were more isolated and vulnerable to abstract processes than in times past.

Indeed, the fight against liberalism (parliaments, market structures), or in psychological terms against pluralization (inclusion, tolerance of multiple moralities, or subjectivities), was predicated on a fear of loss, a fear of emotional repudiation or rootlessness, a fear that derived from the fact that this national world, more and more dominated by the bourgeoisie, really was a more indifferent world. The traditional sorts that Mannheim referred to were willing to trade the right to express independent interests for the right to belong to a community. By wholeheartedly joining and belonging, moreover, they were organizing a way of defining themselves as good and admirable. That is, if inclusion was effective and everyone was entitled to call himself "citizen," or be called that by others, then one could no longer hold on to the feeling of being special and peculiarly worthy and therefore an object of interest and concern. The good are never abandoned, though the bad may be. This nation–people–race argument, in whatever form it may have appeared, was a denial of the possibility of separation and loss and an acceptance of authority, which would not only provide security in a physical sense but in a more profound psychological sense. One could identify and justify the right to consideration only by defining those who were to be excluded and the reasons for that exclusion. Hence, the national and "racial" experience in Europe, and especially the form it took for Jews. Love of one's country, people, or past does not necessarily require hatred of another's; but this was required in Europe, as one cannot think of European nationalism without thinking also of the depreciated and despised other.

Mannheim distinguished conservative thought and action

from liberal, bourgeois, or progressive thought and action in this context, by their orientations to time.[12] The bourgeoisie, according to Mannheim, feed on their "consciousness of the possible," they transcend the given present by seizing on the possibilities for systematic change that had become morally legitimate for them. Bourgeois, liberal, or progressive types are, therefore, future oriented. Conservatives, by contrast, are past-oriented, approaching things in some way "from behind." Everything derives its significance for liberal, progressive thought from something either above or beyond itself, from a vision of the future or a transcendant norm. But conservatives see "all the significance of a thing in what lies *behind* it. . . . Where the progressive uses the future to interpret things, the conservative uses the past."

This distinction between liberal and conservative thought is important, but it is too narrowly cast, considering the historical context outlined above, especially with respect to the analysis of conservatism in terms of a specific ideological and class content. Contemporary conservatives in American society, for example, also seek to retain or recover a valued and threatened past, but it is a past characteristic of the rationalizing bourgeoisie defined by competitive autonomy, individual endeavor, and the inhibition of emotion, an ideological orientation that cuts across class lines.

It is even more important to note, however, that conservatives are not bound to the past only in these terms. The emphasis in Germany on the binding power of the community and the volk, on the repudiation of any level of independent political activity, and on the barrenness of capitalist preoccupations with private interest and personal gain can be understood in the sense Mannheim intended, that is, a version of traditional conservative reactions to liberal–bourgeois society. But the more rigid racist and volkish conceptions represented something else again:[13] The wish to return to or restore something that had existed in pure form, to renew or recover the blood, the devotion to a longing rather than a reality (as exemplified by Himmler's strange identification with Henry the Lion),[14] and the invention of a state in which the past could be recaptured and enjoyed forever, or at least as long as the human mind could conceive, is more consistent with Mircea Eliade's conception of primitive or pagan orientations to time than with Mannheim's conception of conservative orientations to time. Eliade writes that

> the life of archaic man (a life reduced to the repetition of archetypical acts . . .), although it takes place in time, does not record time's irreversibility; in other words [that life] completely ignores what is especially characteristic and decisive in a consciousness of time.[15]

Mannheim's conception of the past-oriented form of conservative thought involves a sense of process, an organic view of change over time, a history, even if it is retrospectively falsified to make it grander and purer than it really was. There was such a conservative view in Germany. But the Nazi view of the past was quite different, it was based on the timelessness of race, on the eternal, not on an organic view of social process over time. The Nazis were seeking to recover the blood, they were not interested in the throne, the traditional class structure, or in any bourgeois or aristocratic conception of hierarchy. The racial Nazis especially could be quite radical in this sense, as with the bigamous morality contemplated by Hitler, Himmler, and Bormann, or their hostility to Christian denominations based largely on their hostility to a religious conception of equality in which servants and masters appear naked and undistinguished before God.

For the radical racial Nazis the present did not gain dignity from the past as such, but only from the eternal. The struggles of life made sense only as evidence of an abiding reality, that is, as the repetition of archaic struggles compelled by the blood and not by historically derived social conflict. Hence the image of the Eternal Jew seeking in his boundless arrogance to ruin Aryan blood. A struggle for the blood, and the space needed for it to flourish, was essential and primary, and everything else secondary. As Hitler put it, anything that is not race in this world is trash.[16]

Hermann Rauschning reported Hitler saying once that he had "to liberate the world from its historical past. The nations are the manifest shape of our past history."[17] It seems entirely plausible that Hitler would make such a statement in the sense that he was profoundly ahistorical. To be sure, he would boast that everything he said and did was historical, but he meant by that memorable, worthy of being recorded and preserved; he was not talking about history as process, as irreversible change over time.[18] Hitler said the Nazis would choose their own way, they would abandon the traditional German thrust to the south and west of Europe and turn to the east.[19] That is, the basis of foreign policy was primarily racial as was the conception of the Greater German Reich. Hitler was after space to guarantee the recovery and renewal of the blood. Hitler and other racist Nazis were past oriented, but they were not interested in any historical considerations that might have animated the monarchy or the bourgeoisie, and the prewar period did not constitute any precedent or establish any obligation for the Nazis.

Orientations to Struggle and Violence

The Nazis held struggle and violence to be an irreducible first principle of life "which is neither derived nor in need of proof," a basic condition of all nature and the basis for all higher development.[20] There is no need to emphasize here how many times it was repeated that war is life, that war is the father of all things, that life is struggle, that nature is an arena of perpetual struggle, that only force rules, compelling the victory of the strong over the weak, that force is the first law.[21] "Those who want to live, let them fight, and those who do not want to fight in this world of eternal struggle do not deserve to live." [22]

Violence was a preferred, exalted mode of behavior, the willingness to fight served as the highest criterion of worthiness for any individual, irrespective of all considerations of caution or personal safety. The mechanistic perfection of the marching columns was meant to control apprehension about self and body, asserting the viability of life through order and providing a basis for mastery of the world. There was a constant dwelling among the Nazis on the front-line experience, on the virtues of war, on the real or anticipated confrontation with pain and death, with fantasies of unlimited accomplishment based on military prowess. The Nazis were constantly preoccupied with military adventures, meaning to establish their superiority through exploitation of the weakness of others, urging aggressive action as if there could be no realistic limitations to such action. One had constantly to prove "the right to exist," in never-ending combat, the ideal being the farmer–warrior of the eastern reaches in Hitler's and Himmler's imaginanation.

The emphasis on struggle and violence served a number of purposes, initially the mobilization of people to resist the humiliation of defeat and the fostering of a spirit of recovery and renewal in the face of a depriving and unjust reality. Hitler seemed to understand that feelings of humiliation, or fears of enforced passivity, are most forcefully mastered by injunctions against appearing shameful and contemptible, independent of any moral norm or obligation, as feelings aroused by fears of weakness, unworthiness, and shame are most compellingly controlled by action that confirms pride and superiority without any regard for its effect on others.

This emphasis then also served as a means of encouraging people to realize a cherished ideal: combative masculine activity in a spirit of discipline and sacrifice. This ideal was tied to a vulgar and inconsistent "Social Darwinism," to ideas of race and blood, and of a menaced future, expressed in pseudoscientific and semiscientific terms that permitted a "realistic" stress on the eternal laws of nature, on "accepting things as they really are," or "seeing the world as it really is," and fighting it out with "brutal determination" in an unyielding acceptance of the need continuously to fight for life, the voluntary sacrifice of all, and the merciless treatment of racial enemies.[23]

The physical courage summoned by the Nazis, among the SA in the fight for the streets, and later among the Waffen SS, the wish to search out and face dangerous and challenging situations, required such an ideal. But it also required a complementary ideal adversary or adverse circumstance. It was more than ironic that the ideal adversary lacked the means, the numbers, and the organization to resist their destruction and the adverse circumstance referred to nothing grander than the tension experienced in the routinized mass murder of defenseless people. The vision of a warrior brotherhood bound by heroic deeds, which animated many of the soldiers of the SS, was betrayed by their own leadership, which bound them forever by nothing loftier or more heroic than the "secret" of the camps.

Orientations to Experience and Affect

The experiential and affective orientation of Nazi ideology is a more complex matter than the orientation to struggle. It is not easy to explain what was meant by these things, although feeling the body as well as feelings as such were involved. This was the most genuinely radical, if confused and contradictory posture to emerge from Nazism, providing some, at any rate, with a sense of the potentiality for a radical breakthrough to new forms of life, allowing the movement to be experienced in some sense as revolutionary. The repudiation of the disciplined self-control of "cold intellect," permitting the domination and not merely the reintegration of affect in political life, was attractive because such strivings are active, they can be gratifying, and they had been suppressed. The posture was inherently contradictory on two counts, however: First, Hitler's accomplishments depended upon such a degree of personal self-containment, such a genuine spirit of indifference to others except as they served him, that the anticipated emotional relationship between Führer and Gefolgschaft could never have been realized.[24] Second, the commitment to technical–scientific rationality (to industry over agriculture, urban life over rural) was indispensable for the military machine that was supposed to guarantee hegemony in Europe and for the real social wealth the Nazis required to ensure domestic stability. The tension between this commitment and the opposed, often preferred, commitment to emotional–instinctual strivings (as exemplified in the peasant's attachment to the soil), was controlled by Hitler personally. There was no way to resolve this tension in structural terms, except as the war and the concentration camps provided an arena in which both could find integrated expression. The camps became in fact the scene of the wildest, most unrestrained sadistic excesses and also of the methodical mastery of the technical–engineering problems presented by the demands for mass murder.

The experiential–affective orientation, which was important for all elements on the right so that many kinds of people could identify with it, was expressed in the most exalted manner, typically as the immediate experiencing of unanalyzable but nevertheless understandable tendencies. Kasimir Edschmid, the expressionist poet writing earlier about the new "poet–man" about to emerge (1919), expressed it as follows:

> His life is regulated not by trivial logic, without any rationalistic or shame-producing morality, but only in accordance with the tremendous calibrations of his heart. . . . He does not think about himself, he experiences himself.[25]

In Nazi usage this orientation referred to feeling oneself, especially in the body, risking danger in combat, for example, so that one could enjoy—or suffer—the physical exertion and the gamut of emotions, a far cry from the staid, despised bourgeois preoccupations with business and home. Thus, heroic man was about to replace economic man, and the emphasis was constantly on force, energy, vitality, restlessness, movement in the body, movement through space—as invariably and favorably contrasted with the bourgeois emphasis on reason, on mechanical, distanced, unfeeling intellect, which sacrifices the body. This ideological orientation to experience and affect was supposed to end the modern slavery to technology and to artificially constructed (emotionally inhibited) societies; man would become natural again, instead of being a mere cog in a network of mechanical functions. Bourgeois society invariably prohibits vital, natural, instinctual processes; the categorical imperative is a negation of racial vigor and communal life, leading as it does in the bourgeois world to self-seeking gain.

There was also a more familiar, pedestrian expression of this orientation, of the following sort: "A National Socialist stance cannot be adopted through reading or thinking but only through experience."[26] Or again:

> National Socialism cannot be apprehended intellectually. Unless you feel in your bones that National Socialist ideology is right you will never be able to grasp it. If, however, your heart leads you to become a National Socialist then gradually, in the course of time, the rational principles of the Party's ideas will become clear."[27]

There were endlessly repeated variations on the idea that race, volk, movement, or leader evoke immediate loyalty not subject to rational or logical criteria but encompassed rather by metalogical criteria as expressed in symbols of fate, heroism, providence, community. The willingness and capacity to experience such feelings of loyalty derived from "sound instincts," or the call of the blood. Eugen Lüthgen told the students the night the books were burned that "the voice of blood speaks louder than the voice of intellect."[28] Instinct was superior to reason and from instinct comes faith: "Instinct is the voice of the blood, the voice of honor, the voice of racial destiny; it has no need of intellectual rules and laws."[29]

The Nazis wanted to organize a social order based on emotional commitments rather than abstract rules, seeking for an uncritically absorbed response to command that would permit immersion in a world of movement, a response that would be heartfelt, derived from "sound instinct," and therefore independent of any critical posture or commitment to self-interest. Emotional commitments would lead to a higher form of communal experience centered on the leadership principle and bound by sentiments of loyalty and honor. The Nazis wanted actively to undermine commitments to objectivity, which was, in their vocabulary, a term of abuse or derision (mania for objectivity, curse of objectivity). All judgments were to be rendered according to their effects on volk and blood and never from the standpoint of abstract reason or legal obligation.[30]

The Nazis repudiated concept formation based on abstract or

inferential criteria and elevated concept formation based on perceptually impelling, rather immediate visual, auditory, and oral experience. This was underscored by Hitler's preference for the spoken word and for the huge, overwhelming, often brilliantly organized political rallies that were meant to loosen cognitive controls, fostering a sense of emotional fusion with a valued community. This is not a bad or dangerous thing in itself; on the contrary, in segregated spheres of Western experience (in religious, artistic, and athletic activity, for example), it is enjoyed, or at least expected and tolerated. But emotional detachment and critical analysis have been the standard for activity in the political and economic spheres, no matter how ill-conceived or poorly realized.[31] Hitler rejected this stance as self-defeating, insisting, among other things, that the oath of loyalty be taken to him personally and not to an office or an abstract idea, refusing to separate the idea from the man who embodied it. He did not anticipate or require critical functions.[32]

This affective orientation to reality, identified also in Hitler's view of himself as the organizer of a spiritual regeneration based on his capacity to divine the volkish soul, was celebrated as a liberating, elevating principle; the consequent behavior was considered superior to abstract reason, the disease of which Germany and the West had to be cured. The Nazi emphasis on "sound instinct" was confirmation of a static, "eternal" reality, while the termination of commitments to cognitive orientations legitimated, at least for some, the instinctual attachment to people and things and a corresponding lack of judgment concerning them. This was the basis for the Hitler cult, the acceptance of his definition of reality, of the sources and purposes of struggle, of his definition of the enemy and how the enemy was to be treated. In any event, one was not supposed to question, that is, examine consciously from a critical standpoint, orders, instructions, interpretations. Those who could not accept this emotionally kept quiet, protecting themselves (and Hitler's precarious self-regard) by not interfering with or contradicting interpretations of events, hoping never to have to make a stand, though some of them finally did so. Those for whom Hitler mediated reality simply believed in his virtues.

Orientations to Hierarchy

The experiential–affective orientation was the necessary precondition for the Nazis' desire to organize a hierarchical social order consistent with their racial and volkish conceptions, especially the leadership principle. Affect and hierarchy together implied that behavior was to remain in the realm of family-like processes, in repudiation of atomistic, competitive, individual strivings imposed by capitalism, but also in repudiation of the Marxist commitment to universal inclusion and the rational mastery of technical structures. Morality was to remain focused in external authority and not in individuals or in self-activating groups.

Hitler always acted as if the integration of impulse and restraint was beyond the capacity of ordinary people. He assumed that people would behave like apes if they had complete freedom of action; and in any event, individual liberty weakens the capacity for order and organization and hence the capacity for struggle.[33] Hitler preferred to accept responsibility, insisting that leadership become responsible in all organizations, distributing the functions of restraint and control, segregating authority from criticism. The leadership principle meant precisely that the hierarchical order could never be reversed nor could subordinate individuals be allowed to assume critical functions.

Hitler had no regard for the traditional exclusive class structure, either in its aristocratic or bourgeois forms. He had been too much the victim of that system, and he had too high a regard for his own capabilities to have any regard for that kind of class exclusiveness. He knew that he could have gone through life unrecognized; he wanted a society open to talent, reflecting his own triumph, which he saw as of the highest value to the community.[34] But when Hitler referred to talent, he meant *racial* talent as expressed in heroic deeds, not in the routine mastery of rationalized technical tasks. There was a broadly conceived fascist version of the warrior brotherhood, men who distinguish themselves in war and are rewarded for their service. This was indeed the end of ascriptive status and the recognition of merit as a basis for reward. But this was also a world exclusively of soldiers devoted to perpetual warfare, the end of bourgeois preoccupations with business and family. This repudiation of bourgeois expectations was even more pronounced among the Nazis, for whom worthiness included racial criteria as well. There is no contradiction between the principles of hierarchy and merit in these terms: Merit was a method of recruitment for a racially exclusive hierarchical order.

The pattern of dominance and submission that was supposed to bind relationships among Germans was extended in special ways to include Nazi relationships to subject and "inferior" peoples. The Nazis meant to become economically invulnerable through conquest. The obvious public concern, the secondary goal that most Germans could identify with, was the avoidance of passive dependence upon the economic good will of others. The primary goal, however, was to ensure that Aryans became the exclusive possessors of everything valuable and enviable in the world. The Nazis would simply take what they wanted as compensation for their unjust treatment and as an expression of their superiority with no sense of gratitude or indebtedness and no regard for limits.[35]

The concept of economic invulnerability—autarchy—was not original with the Nazis. Moreover, the concept was rather a complex one, including a cognitive, a moral, and a wishful content. The Nazis would have liked to create an integrated economic empire, which would not only have solved the problem of blockade but perhaps even the problem of social division in Germany as well, as no German would then ever experience substantial want. It was also strongly felt that Germans *should* have what they deemed necessary because they were vulnerable, they had been historically deprived, and they were culturally superior to the people they meant to exploit. Autarchy conceived in these terms could be regarded as within the bounds of imperialist–chauvinist sentiment as expressed in Europe and elsewhere.

The wishfulness that dominated the conception was distinctive, however, and made Nazi aims peculiar even as measured against Western aims in Asia and Africa, not noted for their generosity. It was the Nazis' sense of grandiose entitlement, their ability to act without restraint, their passion for invulnerability, which all others realized had to be compromised to some degree, their unabashed expressions of racial superiority that justified every excess but could not be established on any grounds satisfactory to conservatives, liberals, or Marxists, the very boundlessness of their claims that proved so alien and threatening that the Western powers and the Soviet Union were compelled to hold to their alliance, no matter what they thought of each other, in order to finish them off.[36]

The assessment was accurate, for behind the economic goals were the racial ones dramatically expressed in unguarded flights of imagination: unlimited prospects for Aryans and utter contempt for all enemies, real and fancied. The Germans would become the richest nation on earth, they would come out of the war "bursting with fat," they would take everything they needed without apology.

> I believe I can say with confidence that after the war we will be the richest nation on earth, and that it will at last be possible for us in our development to solve all the problems which have always frustrated us in the past, because our mental and geographical perspectives were simply too narrow and could not comprehend everything.[37]

In the wider perspective Russia would become Germany's India, but the consequences would be much greater. The SS would be given complete independence, especially in the East, where they would be free to breed a pure race of warriors, the envy of the world. The Spanish, Dutch, Portugese, French, English, they all had great empires but their day was past. The Germans would have the greatest empire in history, and it would never pass, at least as time is reckoned by ordinary standards. Polish and Russian culture would be destroyed; those elements of the population that survived and had not fled beyond the Urals would work for their masters. The Poles and the Russians were to be stripped of everything valuable: machines, art, even children, if they were racially suitable. The Germans would be replete, everyone else would be emptied, not only of goods but of virtue so that they could die and no German need think twice about it.[38] The Nazis would look after their own and no one else in the world, a policy pursued with such ruthless abandon that even Goebbels and Rosenberg considered it a grievous error.[39] The Jews of course would disappear. There may have been arguments between Hans Frank and Heinrich Himmler, Frank insisting that it was necessary to employ skilled Jewish labor in the wartime emergency, Himmler insisting that all Jews must perish regardless, lest they survive, recover, flourish, and seek vengeance. But all this was only a matter of two rival conceptions of the same spirit of entitlement; any glimmer of human sentiment in Frank's position was incidental.

The Germans were masters and they could do and take as they thought necessary or as they pleased. The Nazis kidnapped children of "good racial stock," because wherever Aryan blood appeared it must belong to them. Every such kidnapped child returned to Germany was one more fighter for them, one less for the others. The Nazis would attract Aryan blood to themselves, they would go on "fishing expeditions" for the blood, station racially valuable units in run-down areas to freshen the blood, and whatever could not be rescued would be destroyed. This was passed off as a matter of sober realism: As Himmler said, it may be called cruel, but nature is cruel.[40]

Nazi racial policy was implemented in these terms and all common sense interpretations of Nazism as bourgeois, imperialist, or as naked power organized for its own sake, crumble in the face of this fact. It is true that Hitler made fun of SS mythology, that he did not want Nazism to decline into a form of cultic worship, ending as a substitute religion. In his own view, Nazism was based on realistic, scientific appraisal and did not require mystical preoccupation with ancient stones and bones. It was one thing to elevate Aryan blood to the heights that nature intended, to see to the care of the people who shared the blood, and to foster racial purity. It was quite another on that account to make an issue of Wotan, Widukind, and Henry the Guelph. Hitler respected Greek and Roman antiquity, but Germany had no such monumental past, no glorious ancient culture. Himmler was foolish to be fussing over the archeological remains of primitive Saxon villages; the Greeks were Aryan, after all, and that was enough. Still, Hitler had plans for the SS and for the race of which they were the best representatives, and all Hitler's contempt proves is that there was more than one racial conception among even the SS.[41] Himmler was the policeman, the scourge, ice-cold and duty bound; he would see to the onerous tasks. If the Nazis lost, none of it would matter anyway. And if they won, Hitler's word would be supreme and things would then be settled as they should.

Conceiving the Final Solution

Nazi orientations to the past, struggle, affect, and hierarchy were culturally familiar and sufficiently ambiguous to encompass a heterogeneous population that could interpret these orientations as interest, morality, and wishfulness dictated. But among those for whom Nazism justified grandiose expectations of unlimited accomplishment based on racially animated aggression, recruited at the beginning or along the way, the implications of these orientations are underscored by the use of a particular concrete body language, signifying an incapacity for, or perhaps an indifference to, abstract conceptualization leading to a confusion of categories—for example, inferring good intentions from good physical form, or assuming that people identified by real or potential unity of form must also harbor a unity of feeling about the world.[42]

This concrete body language is manifest in a constant stress on welding, fusing, binding bodies, on organism and immunity, or on rot, decay, decomposition, dissolution, on blood and on pollution or recovery of the blood. Hitler spoke of the Jews within the body of other people, as Nazis spoke of the deep roots of the Party in the body of the people. Walther Darré referred to the peasants as "the biological blood renewal source of the body politic," a fine but by no means unique Nazi turn of phrase.[43]

This concrete body language was exemplified by continuous, unceasing reference to "the blood." This blood reference ("the song of my blood," "blood witness," "blood flag," "blood stream," "the order of good blood," "the botanical garden of German blood," "blood and soil," "intoxicated and luminous blood," "struggle is our blood," "the earth-rooted German's glowing blood," etc.),[44] could be heard abstractly, as metaphor, pertaining to good human nature expressed in simple work and courageous sacrifice for a beloved and intimately known land and people; or it could be heard concretely, as a physical fact, a matter of heredity, which led not only to good physical form but to good moral form as well, so that the people of the blood were not only peculiarly endowed but also peculiarly deserving. Either reference implied a world of the good, the protected, the elevated, as opposed to a world of the bad, the degraded, and the abandoned.* But insofar as the blood reference was taken quite literally, as a matter of the body, it was not only limited, a factor that served to define goodness and community, it had as well a glorious past from which the volk had strayed but to which they could return, a notion that repudiated the fear of irretrievable loss.

The blood reference was also, therefore, a reference to time, and to the extent that the blood could be saved, recovered, improved, or preserved, a sign that the effects of time were reversible, that change that violated the purity of the blood could be halted. Hitler promised to put Germany back on a track abandoned 600 years earlier, and to restore Aryan blood as it had been before its contamination, a result of Jewish guile expressed through Christianity and Bolshevism, confirming to Aryans that no social process was irreversible if their will was firm.

The Nazis were incapable of envisioning a future separate or distinct from the body–biological past, and they were also after timelessness, as time is measured by change. Nazism was characterized by a pseudotemporality, Nazism lacked chronology, a continuous, processual sense of time with its attendant intimations of mortality, the conflicted deterioration of cherished forms, and the discontinuities that illuminate human vulnerability but that must

---

* That the Jews were bad and degraded was evident, as the East European Orthodox and Hasidic Jews were taken as the physical sign of a moral capacity. But that the Jews were also abandoned was not evident and the Nazis went to some lengths to establish the feeling for Jews and for themselves, that is, that Jews were not valued by anyone, they were outside the boundaries of human sympathy and therefore not entitled to consideration or protection.

nevertheless be assimilated in imagination and accepted. Granite would ensure that Nazi monuments lasted forever, Hitler said. "In ten thousand years they'll still be standing, just as they are, unless in the meantime the sea has again covered our plains." [45] If time was proof of the limits of power, then time did not count. Certainly nothing would change as the result of any social process.

The Nazis produced little of abstract intellectual interest in any field, but this was especially true of history. The Nazis were not concerned with history, with life as it changed over time. The Nazis were concerned with the past, to be sure, but as genealogy (the family tree, the clan book, the ancestral office of the SS). The Nazis were searching for the timeless qualities, for what did *not* change, as the blood could be made invulnerable against time and stand as evidence of permanence and indestructibility.

It was in this sense that Robert Ley boasted that the Nazis had "found the path to eternity." [46] The Catholic Church had lasted for 2000 years, the Reich would last forever.[47] Hitler's grandiose architectural schemes, meant to impress outsiders with the grandeur and might of the Reich, were also meant to emphasize a static sense of time. Hitler told Hans Frank that the buildings of Nuremburg would be so gigantic "that even the pyramids will pale before the masses of concrete and colossi of stone that I am erecting here. I am building for eternity."

> Because we believe in the everlasting continuance of this Reich, insofar as we can reckon by human standards, these works too shall be eternal . . . shall satisfy eternal demands. . . . [T]hese buildings . . . are to tower up like the cathedrals of our past into the millenia to come.[48]

This impression of the future, of the all-powerful Reich, served the practical purpose of accommodating people to the need for "detoured wishes," separating people from the present, which promised to be difficult over the short run, while encouraging them to believe in a successful outcome. This was a crucial role for Hitler, encompassed in his ability to appear strong, however that strength was defined, and also in his optimism, as he continuously pronounced his vision of the future, compelling the public and his own colleagues to believe in themselves and their special destiny, to believe, under duress, that there would be a future.[49]

Hitler's ability to focus attention on the future was facilitated by earlier national successes, which reinforced the fantasy of special destiny and the feeling that failure was due to external, alien forces and not to personal or national shortcomings. People were assured that they could live up to ideal standards, that earlier failures to do so were not their fault, that discipline, sacrifice, and the perfectionist traits, so well expressed in the preoccupation with the military calling, would be effective again. Planning was thus concentrated on the future for the sake of morale, but also for the sake of a belief in recovery from a temporary lapse, an unfortunate parting from the eternal. Indeed, the massive effort had no significance, for Hitler at least, except to make things as they ought to have been and once were.

Lived time, or history, was of interest only as it was encapsulated in monuments, stone and concrete proofs of racial viability, the timelessness of race and blood. The everyday life of ordinary people was of no consequence whatever and could be destroyed as it was necessary or useful to do so. Some people, of course, were not viable at all, or represented a threat to the racial viability of others, and everything of theirs could be destroyed without trace and without remorse.[50]

This orientation to the past was indicative of an inability to compromise cherished ideals threatened by social change, particularly as this involved the social acceptance of hitherto despised people and practices. Loyalty to such exclusive ideals, perceived as "natural," mandated looking backward and this interfered with the ability to appraise reality, specifically the meaning of change, the identification of those responsible for change, the interpretation of how change "really" occurred, what the past was like, what the future could be like, definitions of self and others, and so on.

The concrete body language may then be understood in terms of this problematic ability to appraise self and society in a process of change. Jews were described in this context as lacking a stable bodily form, a sign of their divided internal nature. The Jews had no culture of their own and no capacity for creating one because they lacked a coherent center, a culture-forming will. However, the Jews were constantly mingling with other people, interfering with their cultures, seeking to dissolve their bodily form with the expectation of dissolving their cultural form as well. And the Jews had devised all kinds of deceitful ruses to accomplish this aim: Modern art was their invention, rationality their intellectual technique, the city their haunt, the leveling principle of democracy their paramount ideological concern. The Jews fostered equality, which was false on principle and tended toward a racial chaos, which was the precondition of their triumph.

The merely rational mind, trained to dwell on externals, might well ask how it could be that Jews ran the stock market, standing behind all the costly and bewildering financial manipulations that plagued the modern world—and then also the international revolutionary movement dedicated to the destruction of that very market. The answer is clear once one makes up one's mind that the Jews are a "ferment of decomposition," [51] a racial *principle,* a failed and internally divided people seeking to reduce all form to their own chaotic formlessness. For if that was not so, then how explain that one people, and so few at that, could take so many forms, be in so many places, do and think so many contradictory things, and acquire so much influence.

The racist Nazis could not understand otherwise the different forms in which Jews appeared all over Europe: dark and swarthy, blond and blue eyed, clean shaven and bearded, atheist, converted, reformed, Orthodox, Hasidic, occupying no delimited geographical space, belonging to no country, (obviously different from all other people except Gypsies in that regard), but still occupying positions of great social power in many different places and producing all manner of incredibly influential, socially divisive thought (which made them entirely unique and distinctive). This only apparent confusion was not a matter of class or nation, it was a matter of race, of a racial group that lived everywhere, adapted to all conditions, customs, and languages (speaking the language of the people among whom they lived but keeping perhaps two of their own), passing in polite society or resolutely remaining different, but in any case sowing disunity and dissolving form.

Blood or race (heredity, genetics) do not, after all, play tricks: There was an explanation. Jews were compelled by their blood to create a world suitable to their own racial substance and nature ("unfathomable nature") provided them, along with all other parasites, with a capacity for camouflage so tricky that it required experts to see through it. The typical Jew is easily recognized but Jews, like chameleons, can also merge with their surroundings and appear totally un-Jewish. This was so serious that Jews were made to wear the yellow Star of David so that innocent Aryans would not be contaminated by the unwitting contact.[52]

This concrete body image of the Jews, which reflected their fractured nature, was constantly compared to unified Aryan nature. Jews represented cunning, adaptability, dissimulation, yearning for power, and cruelty, which made good their physical lacks. Germans or Aryans represented firm character, genuine emotion, concentration, clarity, inwardness, order.[53] But this Jewish–Aryan polarity was expressed particularly in the terms form–order, formlessness–chaos, and in these terms the Jews did not merely stand for disorder, the disintegration of form: The Jews were the motivating element of all such manifestations, which not only served their purposes but expressed their being.[54]

Any international group that fostered an ideology of inclusion, that failed to make the necessary distinctions of race and blood, was a threat to order, a victim of Jewish contamination, including all at once Christianity, Bolshevism, and freemasonry. The Jews were behind Christianity and Bolshevism especially: Having failed to destroy Aryan blood with one, they now meant to succeed with the other. The Jesuits, for example, were still described as the bacilli of radical antinationalism, a power that must attack unique states. But then, Roman Catholicism had never shaken off its Hebrew origins. Paul (the Oriental rug merchant) was the Jewish destroyer of Germanic ideology, as the Church continued to express a characteristically Jewish–Oriental structure identified by its emphasis on such ideas as revelation.

This illusory seeking after unreal worlds was the result of an "inner infirmity," a "lack of unity and stability," a lack of all capacity for regulated form and order. And it was peculiarly Jewish:

> The nature of the [Jew] is marked by a singular dualism and unrest. The [Jew] lives in a continuous state of tension, alternately flaring up in sudden passion, then abruptly plunging down with a total lack of will and strength. . . . The [Jew] is much too concerned with himself, his insufficiency, his duality, and his inner fragmentation.[55]

The Nazis were so pleased with this analysis of the chaotic, fragmented nature of Jews that they claimed to be able to distinguish the efforts of Jewish intellectuals on the internal evidence of their work alone. It did not escape their attention that Jews were heavily involved in theoretical physics and mathematics while Germans were involved in experimental physics, one focused on vague abstractions, the other on reality. The Jewish mind dwelt on analysis, the German on synthesis, one dissolving, the other organizing reality.[56]

This capacity to collapse all distinctions, to derive all social ills from a single source, is a direct translation and explanation of a social reality, the need in modern societies to absorb different groups continuously and to accommodate and reconcile their peculiar subjective strivings. The criticism of modern society and of the concomitant pluralization of forms, especially by conservatives, is precisely a complaint about splintered subjectivity, a phenomenon that was linked to the social acceptance of Jews. As T. S. Eliot put it, for example, "The population should be homogeneous. . . . What is still more important is unity of religious background; and reasons of race and religion combine to make any large number of freethinking Jews undesirable."[57]

The complaint was directed not only to the proliferation of interests and parties, though that was bad enough, and not only to the proliferation of moralities, as these moralities were focused in individuals and not in society, but especially to the proliferation of every kind of sexual and narcissistic expression tending to the dissolution of form. Atonal music, cubism, futurism, expressionism (the "exotic distortions" of which were specifically related to the Jewish blood of its creators), experimental literature and theatre (either pure form and therefore sterile formalism, in addition to incomprehensible content; or the dissolution of form, and again incomprehensible content), psychoanalysis, cabaret life, corrosive humor that spared nothing, pornography: Berlin in the 1920s and early 1930s certainly, but in a decisive way Hitler's perception of Vienna as well, at an earlier time. Vienna was complicated by racial diversity, the descendants of all the peoples that inhabited the Empire, "and thus it is that everyone receives on a different antenna, and everyone transmits on his own wavelength." Hitler despised the Hapsburg Empire because of the failure to ensure the domination of the German element, which as a result had been swamped. Germany itself was infinitely more worthy but was now also threatened with degeneration. No wonder Hitler complained that the center was not holding, that life was "being torn completely apart," and that the racial problem in Germany and Europe required the most urgent solution.[58]

Many writers have linked the Jews with this dilemma of accommodation to ceaseless change, the need to absorb an increasingly complex reality. The Jews have sometimes been described as victims of forces beyond their control that had shattered their insulated *shtetl* life, but had affected all others as well; and they have sometimes been described as marginal people capable of insightful analysis and criticism as a result of indifference or even hostility to the dominant traditions of the societies in which they lived. The Jews have thus served as a symbol of the lonely, isolated intellectual adrift in an uncertain world; or as a symbol of human vulnerability more broadly conceived; or as a symbol of what is modern, urban, and cosmopolitan as opposed to what is rooted and venerable.

Not this time, however: This time the Jews were perceived as the cause, not merely the result of the process. The Jews did not symbolize, stand for, or represent anything external to themselves, nor did it matter what had happened to them nor did it matter what any of them did. It was rather a matter of the blood, of what they *were*. The splintered subjectivity that was making life perilous was the external visible sign of Jewish nature which, internally fragmented, led finally to the variety of bodily and mental forms in which they appeared and to their desire to organize a social order in which their own peculiar kind could dominate. Every people is true to its kind in a consistent, identifiable, observable way: A German is a German, a Czech is a Czech, an Englishman is an Englishman—but the Jews could belong to any of these national groups and a hundred others besides. The Russians and Poles were inferior and as such they could be dominated. But they were consistent within themselves, true to their kind. The Jews were true too, of course, but that meant they were chaotic, as gifted in fomenting disorder as Germans were in imposing order.

Hitler's friend and mentor, Dietrich Eckhart, wrote that "the Jewish question is the chief problem of humanity, in which, indeed, every one of the other problems is contained. Nothing on earth could remain darkened if one could throw light on the secret of [the Jews]." [59] This "secret" Eckhart and Hitler considered themselves to have penetrated: It was a matter of Jewish character, a masked but unified principle defined by formlessness. The Jews were internally fragmented, they were all alike in this regard no matter how they appear on the outside, and they seek to reproduce a world in which this internal fragmentation could be expressed; they have always sought it, only now they were more successful than ever and coming dangerously close to their goal in Germany.

Hence the compulsive dwelling on the chaotic muddling of the races, the bastardization of the races, blood poisoning, racial death, racial swamp, racial chaos, human flood, human quicksand, Asiatic chaos, Eastern chaos, Bolshevik chaos. Hence also the dwelling on the will to form in Aryan character, and the expressed fear of becoming a formless mass without inner cohesion, a fear now overcome by a unifying ideology and a world–historical leader. The chaotic age was over when the instinct for order was abused; the fragmentation of society was forestalled and the true Aryan spirit manifest in the willing subordination of personal strivings to the requirements of discipline and order could be expressed. Robert Ley told the first class of *Ordensjunkers* that when he looked at them he knew "that the principles on which we mustered you are sound. Externally you already look alike and in a short time you will be alike internally as well." [60]

The Nazis sought after what they called an "integrated character," or what the Italian Fascist, Marinetti, called Hitler's "photographic static." [61] They meant by this concrete and specific physical homogeneity from which willing submission would follow, as character was a reflection of body type. This was not the mechanical form and order of rationalizing lawyers and clerks, superficial and objective, but the organic form, which comes from the blood and is expressed in a spirit of unity symbolized by the marching columns. The Germans had mastered discipline, order, obedience, and this was reflected most clearly in the soldierly spirit. There was no longer any need for private reflections, personal standards, or sterile discussions, nor for any further absorption of alien elements.

The Germans were good material: stable, stolid, loyal, dependable, upright, on their best form, blond, blue eyed, and lean. This meant, too, that the Germans were resolute, courageous, and innocent. If other suitable Aryan elements were added, one type could be derived from such blood, which must be preserved, improved, increased—and could be if the Germans and others but had the will. It was not unreasonable, then, to save this kind, responsible for all creativity, by destroying an alien, hostile breed, down to the last one. A surgical operation was needed, not to become (ostensibly) the occasion for sadistic self-indulgence, but disciplined and merciless, for mercy in the matter of the blood was treason. On this logic Hitler stood Terrence on his head, as if to declare, "I am such a man that nothing alien is human to me."

The concrete body language can be read and translated, therefore, and all the easier too as the gap between its cognitive utility and evident wishfulness is unbridgeable. All the head measuring, studies of facial characteristics, the search for pure Aryan blood, and the remarkable presumption that good moral form would follow automatically from good physical form, confirm this. The body language could have been understood as metaphor or as a literal rendering of reality, but in either case it was an integrating, motivating force, an actuating belief. In either case, too, the language can be taken to express fears of vulnerability, weakness, and passivity in the face of power, and an affirmation that the Germans were as good and as brave as they had been led to believe. Indeed, the racists' purpose in concentrating so much anger on the Jews was to magnify their power, which served not only to explain defeat, but also to enhance the racists' image, as if they could think highly of themselves only by being the presumed target of some malevolent force that they had courageously defeated.[62]

The genocidal behavior of the Nazis followed from such fantasy images of perfect form and chaotic formlessness, ideal images of light and dark, creativity and destructiveness. Of course, there remains the question of whether Hitler ordered the destruction of the Jews before or after he realized he would be defeated, or indeed whether he had ordered it at all, that is, whether the SS had not initiated the genocide on their own, knowing that Hitler would approve of their actions. Whatever the answer to this question is, whatever the sequence of events, the fact remains that Hitler knew about the genocide, he had the authority to stop it, but he was, as Goebbels claimed, the undeviating champion of "the radical solution."[63]

In short, whatever the details of the matter, the genocidal behavior is accounted for by the kind of fantasy thinking I have described. This is a difficult thought; people are in the habit of distinguishing the "real world" from the "fantasy world," as if to say that triumphant and fearful fantasies are not real, which they know perfectly well is untrue. The content of any individual's fantasy is also so remote from actualization that it is assumed to be unique, a psychotic product, in effect, not amenable to systematic analysis and especially unrelated to the social world, two other thoughts which are untrue. Finally, everyone realizes that while fantasy thought is exaggerated in its perfection and imperfection, cognitively disciplined thought is transformed in light of realistic considerations and rendered life-size.

The Nazis' fantasy images were acted on, however, just because the real constraints had been removed. Hitler had patiently and determinedly destroyed or rendered ineffective potential sources of opposition, and he succeeded in establishing himself as the unique and exclusive source of authority. The racist images were systematically organized as a belief, and while not everyone believed, the images still were not contradicted by any authoritative, organized force from within. On the contrary, belief and action were facilitated by a highly effective propaganda and military machine that worked to destroy an already fragmented, disparate, and fragile Jewish social organization, and by virtually universal indifference, if not hostility. Many people, including Nazis, thought that Hitler's racial preoccupations were peculiar, but they either did not take them seriously or condoned them because the secondary goals were worth the risk. In any event, no one contradicted the leader on this issue.*

The racial Nazis feared being undermined and destroyed by a hated enemy, a fear that they tried to control by creating for that enemy the chaotic (psychotic) world they thought was being prepared for them. The Nazis displayed self-glorification and self-righteousness, which in itself is not that unusual. But they did so, and Hitler especially, in the context of a concrete body language that signified an inability to see the world in abstract or relative terms, so that the enemy could be treated with an intensity of anger out of all proportion to any offense they could possibly have committed. But the remarkable feats of violence nevertheless failed to abate anxiety over penetration, contamination, and degeneration by World Jewry, a highly organized, demonic force. For no matter how many Jews they killed, the world still could not be brought into line with their wishes, and if that was so, if people resisted, then the Jews were still powerful and, relative to Germany's growing weakness, perhaps more powerful than ever. The killing in principle was endless.[64]

The boxcars and the camps confirm the story, first because they were meant to prove that no matter what Jews were like on the outside, they were really all alike on the inside. The inner unity was compelled, as all distinctions of class, status, region,

---

* The principal question, after all, is not where the impulse came from, because at the level of impulse anyone is capable of this kind of behavior. The principal question is what happened to the restraints.

religion, occupation, education, age, sex, and language were collapsed, so that only the blood remained; and second, because the expressed fears of contamination, the ritual cleansing and purification, the undisguised ritual devouring and elimination of the powerful, hated enemy in the death camps was as literal and concrete an expression of a distorted image as one can imagine. The fantasy was hardly masked: The camps were referred to as the asshole of the world and the SS were known to greet their victims with shouts of "Hip, hip, hooray, Jewshit up the chimney." People were indeed taken in as animate life and expelled as inanimate waste.[65] *

The Nazis had turned a problem of social conflict, which they could not assimilate in mind or accommodate to memory, into a physical problem of nature and the body; and they turned the central problem of memory, time, into a physical problem of space.[66] The Nazis sought to dominate social conflict by holding the mind static ("Believe, obey, fight"), encouraging movement only in the body; and by holding time static, urging movement only in space, so that if change occurred it occurred only in the concrete, physical dimension, which they could master by technological achievement and by the unrestrained expression of masculine ruthlessness. Space was in fact an appropriate metaphor for what had become the psychological reality, the essence of which was constant pressure against limits and boundaries. There were already only a few psychological limits on the treatment of Jews, Slavs, and Gypsies, and the Germans were constantly being urged to test the external restraints, to see how far they could go. Whether there were any limits in space at all then depended upon the willingness of others to resist, which Hitler allowed himself

---

* It is just not possible to give one group so much power over another without instinctualizing the environment, as no reality considerations intervene between the sexual and aggressive fantasies that must be aroused and the ability to act on them. The victims in fact get beaten on two counts: first by those who enjoy beating, and then by those who are frightened by the unanticipated feelings, for which they hold the victims responsible, of course. Potential victims should always keep in mind the possibility of struggle: If nothing else it might make the executioner feel better.

to think was a matter of the blood and that the better blood was on his side.

Social change exposes the human vulnerability to time, as defeat exposes the vulnerability to humiliation. The emphasis on mind–time underscores these dilemmas; the emphasis on body–space is meant to deny them. The Nazis acted in a concrete, physical, nonabstract world of body–space, preoccupied with thoughts of mastery and autarchy, their orientation to time revealing hostility to change, their orientation to space revealing a desire for invulnerability. The drive to world conquest was the perfect expression of the omnipotent wish to have no need to need.

The Nazi insistence on force, the constant movement in space, the marching and the fighting, served not only to impress observers, but to integrate the participants, fostering the illusion of unlimited achievement through violence and the possibility of control through power. This constant resort to force, the constant stress on struggle and violence, was an attempt to deny the contradictory nature of the human condition and to solve an insoluble problem, the dilemma of loss experienced in time. Hitler especially refused to concede the inevitability of loss. Hence the feeling on every side, the Nazis' included, that there could have been no end to it, not as long as he was alive anyway.[67]

Leadership and ideology were decisive for Nazi practice; and understanding different commitments to leadership and ideology provides the only insight we can have into the motivations for systematic mass murder. Obviously I could formulate these commitments only in general terms, directing attention to significant and often-repeated contents without implying any perfect correspondence with reality, particularly in the sense that everyone felt similarly about leadership and ideology, or understood them in a similar way. In fact, aside from various biographies that indicate that thought and behavior were relatively consistent with what I have described, suggesting that in a general way the position is correct, it is impossible to state in precise terms why anyone acted one way or another. Certainly not everyone who used

Nazi language or who worked to realize Nazi aims believed in what he was doing.[68] Our knowledge of this problem must forever remain tentative and limited, a result of the kind of subjective data that is required to establish such knowledge.

If the manifest failure of cognitive processes to control perceptions of reality is taken as a valid criterion, Nazism can be described as "irrational," a judgment that can hold for the behavior of Hitler and some of his colleagues. I mean by this that no information, no form of reassurance from authoritative persons, no technique of reality testing, no investigation of any sort by any person could dissuade Hitler from his racist beliefs in a Jewish world conspiracy or from the decisions he was compelled to make in these terms. It is not that Hitler decided to act on these premises although he knew better or differently, or that he was manipulating people by taking advantage of their fears, or that he was processing all the relevant information that he might have, considering the gravity and enormity of the decision. The process rather was out of control. Besides, Hitler was so segregated from criticism and so exalted by the people around him that he was constantly being confirmed in what must be judged his fantasy.

It is important to note, however, that in addition to those people who wanted to follow Hitler through to whatever conclusions his sense of entitlement and grandiosity suggested, as a result of some similar or complementary racist vision, there were those who continued to see primarily the moral significance of the National Socialist movement, who took for granted the superiority and merit of the soldierly life, for example, and hoped to achieve a higher type of culture through struggle. There were also those who remained cognitively oriented, continuing to flourish by dissociating themselves from the larger significance of events, by concentrating on personal success and technical mastery in industry, science, or the military. None of these people were in a hurry to draw the line on how far they would follow Hitler, particularly as long as he was successful. But some of these people did finally draw the line; not everyone in Germany was willing to follow Hitler in every step, not even his closest collaborators. Speer drew the line on the destruction of Germany, and he got away with it. Others risked death by refusing to follow orders—

for the destruction of Paris, for the destruction of the art treasures at Alt Ausee; and regular army and SS officers refused to comply with orders on any number of occasions. Afterwards the Nazis routinely said they were "following orders," as if that was all anyone ever did. And their enemies needed to believe that, because Nazism seemed to make sense only as slavish adherence to authority. But many of these people were capable of making discriminating judgments and risking their lives for the sake of their judgments—if the issue was important enough for them to do so. It was the miserable fate of Jews, Slavs, Gypsies, and others not to have been important enough. This decision on the part of people who had the capacity to choose and the courage to act and did not is the real measure of the Nazi disaster, and not any presumed German "sickness" or national character that blunted that capacity or rendered it inaccessible to imagination.

Notes

1. Hans Frank, *Im Angesicht des Galgens* (Munich, 1953), pp. 184, 334. Frank wrote that National Socialism simply became "what so-and-so says or does, by which the representative who happened to be speaking meant himself," and this gradually "replaced the assumptions of the party program. Fundamentally there were as many 'National Socialisms' as there were leaders."

2. These different views of the Nazis reflect the need to reduce the shocking and the bizarre to common sense terms. But Hitler did not mean to create a common sense world and such terms are inapplicable.

3. I am equating internalized morality or superego standards with the concept of ideology because, in the sense that Marx and Engels used it, the concept involves unconscious thinking and involves a process of providing people with moral legitimations for interest-oriented activity.

4. Marxism is organized to prevent reflection, to prevent the individual from seeing himself as an object, or from introspecting on motives. Psychoanalysis compels such introspection, the only basis on which "conscious" activity, such as Marx anticipated, can occur.

5. These are precisely the circumstances in which charismatic leaders appear, serving to establish an interpersonal tie with themselves as the representatives of a restored or novel morality.

6. This conclusion is the first step in the necessary evaluation of Marxism as ideology, in the sense that Marx intended.

7. Ideology occurs everywhere in a high level version (all men are created equal, competition is natural, property is sacred, individual rights are inalienable), and a low level version (accidents happen, he's not bad, he's sick, that's the price of progress, got to compromise, win a few, lose a few, that's life). The Nazis, of course, tried to develop their own high and low-level versions of Nazi ideology to serve as internalized morality.

8. It is important to emphasize again that language is conventionalized by a community of speakers who hold the meanings to be invariant; but what is (relatively) invariant is behavior, not meanings, which are quite variable, as psychoanalytic practice discloses. The common sense view, and one that is also employed routinely and inappropriately by theorists, is that one can infer shared meanings from shared behavior. That is, people learn to accept a common sense world in which variation of meaning is rarely examined because the behavior is appropriate to the situation. Meaning is plumbed in psychoanalytic therapy because the appropriateness of behavior has become problematic.

9. The ability of Catholics to resist Nazism and the stability of their political organizations and their electorate has already been commented on. See Chapter 2, n. 4. See also, for example, Martin Broszat, "National Socialism, Its Social Base," in E. J. Feuchtwanger, ed., *Upheaval and Continuity: A Century of German History* (London, 1973), p. 144.

10. This of course was deliberately manipulated: "The only form in which the SA appears to the public is the closed formation. This is . . . one of the most powerful forms of propaganda. The sight of a large number of inwardly and outwardly calm, disciplined men, whose total will to fight may be unequivocally seen or sensed, makes the most profound impression on every German and speaks to his heart a more convincing and inspiring language than writing, speech and logic can ever do." Joachim C. Fest, *The Face of the Third Reich*, trans. Michael Bullock (London, 1970), p. 142.

11. Karl Mannheim, "Conservative Thought," in his *Essays in Sociology and Social Psychology* (New York, 1953), pp. 88–111.

12. There were other criteria for making such distinctions; for example, the conservatives seized upon the themes of romantic literature and especially the identification of the irrational and the preoccupation with feeling as proof against the rationalizing tendencies of the bourgeoisie.

13. To return to the theme of the first chapter: Wilhelm Stapel said in a radio broadcast in 1932 that the Jews corroded the national community by advocating the outlawing of war when all national-minded Germans knew that war was the father of all things. Stapel also observed that man's soul withered once he immersed himself in the world of the intellect. Cited in Werner E. Mosse, ed., *Entscheidungsjahr 1932: zur Judenfrage in der Endphase der Weimarer Republik* (Tübingen, 1965), p. 519; and in Hermann Glaser, *The Cultural Roots of National Socialism*, trans. Ernst A. Menze (London, 1978), p. 99. These were phrases and ideas Hitler used all the time—as did Ernst Jünger, for example. But they never meant the same thing

by them. Edgar Jung, a proper conservative nationalist, wrote that the state must be an aristocracy, the rule of the best in the highest and final sense. But Jung could not make his version of hierarchy prevail—he was murdered during the Röhm purge. See Walter Struve, *Elites against Democracy: Leadership Ideals in Bourgeois Political Thought in Germany, 1890–1933* (Princeton, N.J., 1973), pp. 317–352. Alfred Rosenberg wrote an immediate response to the court's verdict on the Potempa murders: "Man is not equal to man; deed is not equal to deed. Five Germans are to be shot because [a Polish Bolshevik] has been slain. Justice of this kind defies the nation's most elementary instinct of self-preservation. National Socialism, however, knows no equality between souls, no equality between men . . . no object save a German nation of strong men." Alfred Rosenberg, *Blut und Ehre* (Munich, 1939), pp. 71-73. Rosenberg's conception of hierarchy was racial, Jünger's soldierly, Jung's Christian. But even Rosenberg's racism would not have led him to Hitler's conclusions, or Himmler's. Hitler could say, "All life is bound up in three theses: struggle is the father of all things, virtue lies in the blood, leadership is primary." But there is no way to say what people made of that statement or what they thought he meant, though the decisive distinctions were being and doing, concrete and abstract. People could misunderstand at these levels without any difficulty at all.

If language may be taken to evolve from a concrete medium of primal connections (sexuality, birth–death, body parts, kinship) to an ever increasing associative symbolism, then we have another criterion by which to judge where we are. What Freud tried to explain was that he heard the older layer of concrete representations (money = shit) all the time in psychoanalysis, either directly or in such an unguarded way that it was not hard to infer. And what we have here is a similar kind of *publicly* expressed concrete representation—though of course people typically infer metaphorical import from such representations. See Sheldon Bach, "Narcissistic States of Consciousness," *International Journal of Psychoanalysis,* vol. 58, no. 2 (1977):209–233.

14. Josef Ackermann, *Heinrich Himmler als Ideologe* (Göttingen, 1970), pp. 60–61.

15. Mircea Eliade, *Cosmos and History: The Myth of the Eternal Return* (New York, 1959), pp. 85–86. As Eliade noted, time does not exist for archaic man if he does not pay attention to it; and when time becomes perceptible, when man deviates from purity and falls into duration, time can nevertheless be annulled by restoring or returning to the original purity.

16. *Mein Kampf,* trans. Ralph Manheim (London, 1969), p. 269.

17. Hermann Rauschning, *Hitler Speaks* (London, 1939), pp. 229, 121.

18. Victor Klemperer, *LTI: Notizbuch eines Philologen* (Berlin, 1949), p. 51.

19. *Mein Kampf,* pp. 598–609; *Hitler's Secret Conversations,* trans. R. H. Stevens and Norman Cameron (New York, 1953), p. 436.

20. Wendula Dahle, *Der Einsatz einer Wissenschaft* (Bonn, 1969), pp. 30–33.

21. " A permanent state of war on the Eastern front will help to form a sound race of men . . ." *Hitler's Secret Conversations,* p. 34.

22. *Mein Kampf,* p. 262. Rauschning, *Hitler Speaks,* p. 89.

23. Merciless treatment of the enemy involved the projection of bad feelings and traits onto others (who were described as greedy, cunning, malicious, power-mad, destructive by nature) and then "fighting" them as representatives of a dangerous world. The realistic stress on cruelty was justified by reference to nature and repeated many times. See, for example, *Hitler's Secret Conversations,* p. 166; or Melita Maschmann, *Account Rendered: A Dossier on My Former Self* (New York, 1965), p. 122, where the exploitation and destruction of the Poles, "weakening their national substance," was described to her as "realpolitik."

24. According to Hitler, "the people expect not only that their leaders should govern them, but also that they should look after them." *Hitler's Secret Conversations,* p. 391. But Hitler viewed people as appendages to racial germ plasm and he meant to pursue racial ideals regardless of the cost to them. He could not have acted on such a principle except in these terms.

25. Kasimir Edschmid's "Expressionist Manifesto," quoted in Ronald Grey, *The German Tradition in Literature, 1871–1945* (Cambridge, 1965), p. 50.

26. Michael Schmaus, quoted in Rolf Seeliger, ed., *Braune Universität: Deutsche Hochschullehrer gestern und heute,* 6 vols., (Munich, 1966), 1: 61. It was asserted, for example, that *Mein Kampf* could be experienced but not logically analyzed. Klemperer, *LTI,* p. 242. George Mosse, *The Crisis of German Ideology* (New York, 1964), p. 226.

27. Michael Kater, *Studentenschaft und Rechtsradikalismus in Deutschland 1918–1933* (Hamburg, 1975), p. 184.

28. Paul E. Kahle, *Die Bonner Universität vor und während der Nazi-Zeit (1923–1939),* Wiener Library document, n.d., p. 6.

29. Quoted in Josef Wulf, *Literatur und Dichtung im Dritten Reich: Eine Dokumentation* (Gütterloh, 1963), p. 128. The idea that faith springs from instinct is Hitler's as cited by Rauschning, *Hitler Speaks,* pp. 115, 184, 222. Hitler explained to Rauschning that mistakes made out of enthusiasm could be recovered. "But no one would ever attain a conception of the greatness of our task by reasoning alone; it has to be felt and experienced." Hitler was afraid that men would be talked out of their instinctual capacities, the blood would be sacrificed to intellect.

30. In the words of Hans Schemm, Nazi minister of education in Bavaria, "Intelligence, what does that include? Logic, calculation, speculation, banks, stock exchange, interest, dividends, capitalism, career, profiteering, usury, Marxism, Bolshevism, crooks, thieves." Quoted in Kurt Zentner, *Geschichte des Dritten Reiches* (Munich, 1965), p. 347.

31. This commitment to affect meant, for example, that women were peculiarly possessed of what the Nazis thought were real virtues: intuition, instinct, the binding power of emotional expression. But of course women were to remain subordinate in a culture devoted to war and struggle.

32. Once creativity is defined by racial criteria and its form decreed from above, there is no need for a critical function, since that kind of independence of judgment does not by itself lead to any positive contribution. See Josef Wulf, *Die Bildenden Künste im Dritten Reich: Eine Dokumentation* (Gütersloh, 1963), pp. 124–26.

33. For example, *Mein Kampf*, pp. 402–405. On hierarchy and value see Wulf, *Literatur und Dichtung*, pp. 171–72, 290–92, especially the statement by Hans Grimm, p. 292. This was consistent with Hitler's feeling that people crave domination and feel abandoned when they have freedom. "Obviously, then, those in authority must never permit their decisions to be criticized by those subordinate to them. The people themselves have never claimed such a right." *Hitler's Secret Conversations*, pp. 343, 391.

34. "It's a great time when an entirely unknown man can set out to conquer a nation, and when after 15 years of struggle he can become, in effect, the head of his people." *Hitler's Secret Conversations*, p. 180. Of course, this was but another expression of the anonymous soldier–world leader theme.

35. The German experience of defeat and humiliation created a situation in which the victimization of others seemed appropriate. The Nazis came to power in the first place because they promised not only to resist any further assaults on the particularities of the German people but to elevate and impose those particularities. The possibility of a universal community, which Marx saw stemming from real market forces and class consciousness, meant that people everywhere would have to give up certain individual and national particularities, referring primarily to a distinctive, separate, and competitive self and society. But these were not the particularities the Nazis wanted to impose, which were rather the exceptional particularities of race, blood, volk, worthy of unlimited enhancement.

36. The wishfulness bears the characteristic marks of a narcissistic fantasy: to be superior, aloof, and alone, master of one's fate, forcing others— through power, wealth, even promises of respect—to admire the stance while conceding the wishfulness, or wresting the concession from them, if it comes to that.

37. The quote is from a speech of Goebbels' of February 28, 1942, quoted in Marcus Stuart Phillips, "The German Film Industry and the New Order," Peter D. Stachura, ed., *The Shaping of the Nazi State* (London, 1978), p. 263. The other statements are Hitler's from *Hitler's Secret Conversations*, pp. 506–508.

38. Martin Bormann: "The Slavs are to work for us. Insofar as we do not need them they may die." There are so many such statements that one

could easily become desensitized to their importance. See for example Hans-Adolf Jacobsen, "The Kommissarbefehl and Mass Executions of Soviet Russian Prisoners of War," in Hans Buchheim et al., *Anatomy of the SS State* (New York, 1970), pp. 507–523. The necessary documentation is provided in the German edition, *Anatomie des SS–Staates*, 2 vols. Olten and Freiburg/Br., 1965), 2:198–279. The image created by the Nazis was that of a world of ideally good and bad children in which the good may experience the gratification of wishes without limit while the bad may suffer whatever fate befalls them, because they are unworthy anyway. The virtuous are raised, the bad are abandoned. On the subject of the destruction of Bolshevism and the Jews (the biological basis for Bolshevism) and the conquest of natural resources to guarantee the future of the Reich, see Andreas Hillgruber, *Hitlers Strategie: Politik und Kriegführung 1940–1941* (Frankfurt a/M., 1965), pp. 516–535. The important thing about Hitler's fantasies of the good life is that they included perpetual warfare.

39. *Hitler's Secret Conversations*, pp. 5, 16, 29, 57, 138; on the subject of the Jews, pp. 56–57, 65, 72, 96–97, 236.

40. Himmler: "I really intend to take German blood from wherever it is to be found in the world, to rob it and to steal it wherever I can." "We must take it for ourselves and the others must have none." These and other such statements are used as chapter head notes by Clarissa Henry and Marc Hillel, *Children of the SS*, trans. Eric Mosbacher (London, 1976). See also Ackermann, *Heinrich Himmler als Ideologe*, pp. 207, 209; or Walther Hofer, *Der Nationalsozialismus: Dokumente 1933–1945* (Frankfurt a/M., 1957), pp. 112–113. The reference to freshening the blood by stationing racially valuable units in run-down areas is Hitler's, from *Hitler's Secret Conversations*, p. 352. Hitler was a vegetarian, and that is not hard to imagine.

41. On differing ideological conceptions among the SS, see John Steiner, *Power Politics and Social Change* (Atlantic Highlands, N.J., 1977), pp. 87, 92, 99, 113, 122. On Nazism as science, on the concern with metaphysical digressions, on sheltering the Party from religion, etc., *Hitler's Secret Conversations*, pp. 49, 51, 57.

42. On the confusion of physical and moral characteristics, see Steiner, *Power Politics and Social Change*, p. 234, n. 54. George Mosse has pointed out that the body–soul confusion was a minor convention in racist thinking in Germany over time (*The Crisis of German Ideology*, pp. 70, 88–89, 104). But there is a great difference between the random appearance of a thought in different individuals and the elevation of a thought to such ideological dignity that it becomes a matter of national policy to realize it.

43. The body references are endless. Robert G. L. Waite, *The Psychopathic God* (New York, 1977), pp. 360, 362. Waite referred to Hitler's statement that "An attempt to restore the border of 1914 would lead to a further bleeding of our national body"; and "The Polish Corridor . . . is like a strip of flesh cut from our body." The reference to the Jews in the

body of other people is in *Mein Kampf*, p. 279. On Walther Darré see Karl Saller, *Die Rassenlehre des Nationalsozialismus in Wissenschaft und Propaganda* (Darmetadt, 1961), pp. 107–108.

44. The blood references are also endless. See Fest, *The Face of the Third Reich*, pp. 122, 168, 218, 232, 267, 301. Robert G. L. Waite refers to Hitler's concept of *blutkitt*, "blood cement," and nothing could be more concrete than that. (*The Psychopathic God*, pp. 22–23). The botanical garden of German blood is Himmler's reference to the German colonization of the East. Ackermann, *Heinrich Himmler als Ideologe*, p. 295. Hans-Jochen Gamm, *Der braune Kult* (Hamburg, 1962), for example, pp. 141–147; Klaus Vondung, *Magie und Manipulation* (Göttingen, 1971), pp. 84, 173, 176.

45. *Hitler's Secret Conversations*, p. 67. The "hard men," those square-jawed blocks of concrete that turned up in Nazi poster art, also belong here: Such men would never crumble or dissolve either. If all wishes are gratified, time as the experience of change ceases to exist. The incessant ritual observations were also meant to recall the eternal, of course.

46. Quoted in Klemperer, *LTI*, p. 120

47. Goebbels, quoted in Gamm, *Der braune Kult*, p. 33; see also pp. 37, 39. Goebbels could easily have said such things without believing a word of it. He knew what was wanted, however, and his language was not coincidental.

48. Frank, *Im Angesicht des Galgens*, pp. 320–21, 311. Karl Heyer, *Der Staat als Werkzeug des Bösen: Der Nationalsozialismus und das Schicksal des deutschen Volkes* (Stuttgart, 1965), pp. 65–66.

49. All the speeches, marches, rallies, ceremonies, may be construed as the ritualization of optimism, about which Hitler was quite explicit. One of the reasons Hitler did not like religions was that they were pessimistic about life in this world. Rauschning, *Hitler Speaks*, pp. 262–63.

50. By focusing on what people are rather than on what they do, goodness can be defined by blood and not by deeds or definable types of conduct. Moreover, if values are eternal, then differences, discontinuities, and setbacks are also explained by the categories of the racial myth. Race counts, and not society, so that if Jews and Slavs are different, they are hopelessly different. All the criteria of discontinuity, from empire to republic, victor to vanquished, German to cosmopolitan, rural to urban are explained by the one idea.

51. "Ferment of decomposition" is probably the best example of a phrase pertaining to Jews that could be understood in abstract or concrete terms, or in terms of what people do as opposed to what they are.

52. Saller, *Die Rassenlehre des Nationalsozialismus*, p. 129. From this standpoint the Jews were a racial mixture combining Asiatic, African, and European characteristics so that Jewish individuals could acquire the facial and bodily characteristics of non-Jewish people. There was a considerable discussion of whether the Jews were a mixed or a pure race, whether they

had been pure and become mixed, or had been mixed and become pure, etc. Of course, whatever the answer, the result was the same. Hitler said that the Jews were the most weather proof of people; they could live anywhere, in Lapland or the tropics, unlike the Germans, who needed the sun, though not the tropical sun. *Hitler's Secret Conversations,* pp. 393, 322.

53. On the language used to discuss or describe Jews, see Klemperer, *LTI,* pp. 188–89; Wulf, *Die Bildenden Künste,* p. 145; Cassie Michaelis, Heinz Michaelis, and W. D. Somin, *Die braune Kultur* (Zurich, 1934), p. 77.

54. The terms *order* and *chaos* were actually a German convention, employed not only on the right. Franz Schoenberner wrote of the dark, demonic force of the East, the irrational force destroying Western culture. But this time it was Dostoevsky who stood for this tendency and it was rationalism that was being destroyed. The introduction to the notorious SS pamphlet, *Der Untermensch,* spoke of a "cruel chaos of wild, unbridled passions . . .," an unbridled will to destruction represented by Bolshevism, Jews, Asiatic hordes, etc. Ackerman, *Heinrich Himmler als Ideologe,* p. 212. Hofer, *Der Nationalsozialismus: Dokumente,* p. 280. Robert Taylor, *The Word in Stone* (New York, 1976), p. 182. "Throughout the literature on architecture, there runs a concern for order as a desirable quality in German society . . . The writings on architecture, especially on military buildings, are filled with this concern for order, a concern which goes beyond the striving for good form."

55. Matthias Ziegler, *Illusion oder Wirklichkeit? Offenbarungsdenken und mythischer Glaube* (Munich, 1939), pp. 7–8.

56. This is how Gottfried Benn got into trouble. But see in addition the comments on the critic Alfred Kerr and the writer Reinhold Conrad Muschler in Wulf, *Literatur und Dichtung,* pp. 270, 142, 98–100. On the inner conflicts caused by corrupted blood and the ability to determine moral failure from style, see for example, Dahle, *Der Einsatz einer Wissenschaft,* pp. 223, 235, 164. The quality of Jewish unrest comes from their inner division. But German unrest ("creative power of German unrest") is faustian, eternally young, never satisfied, and always renewing itself. It should perhaps be recalled that Himmler and Hitler thought about Reinhard Heydrich in these terms; Heydrich was a man of mixed race who resolved his inner fragmentation by purely intellectual means and was constantly suffering as a result. Felix Kersten, *The Kersten Memoirs,* trans. Constantine Fitzgibbon and James Oliver (London, 1956), pp. 97–98. Steiner, *Power Politics and Social Change,* pp. 258–59, n. 185. Hitler on the dismissal of Bismarck by Wilhelm II: "The irresponsibility of that young man is beyond comprehension . . . In his whole attitude the heritage of his Jewish ancestry comes out in the completely cynical lack of self-control, which was characteristic of him." *Hitler's Secret Conversations,* p. 526. Hitler thought that all people suffered from mixed or corrupted blood to some degree, a situation he meant to change. Rauschning, *Hitler Speaks,* p. 227. See also the statements of Hermann Franz Gerhard Starke, in Seeliger, ed., *Braune Universität,* 5:70.

The Nazi preoccupation with deriving mental configurations from body types was confirmed by the psychological experiments of Professor E. R. Jaensch of Marburg University. Jaensch concluded that there are two basic types of people, integrative and disintegrative. The Jew is a disintegrative type, though a number of Aryans are too, a result of racial mixing. The Jew, however, is a disintegrative type in pure form, a biological phenomenon, and Jews behave inappropriately even when they do not intend to consciously. Inherently unstable biological processes cause unstable perception, which leads further to unstable political, ethical, and logical conclusions: The disintegrative type tends to be a political liberal. Such a defect, however, sharpens the peculiar intelligence of such a type, and this, in addition to their instability of ideals and their biologically caused adaptability to various conditions, encourages them to achieve economic, political, and scientific leadership. Intelligence testing is a result of their interest and is organized to conform with their own traits; such testing therefore does indicate superiority to Aryans. Such a type is subject to distraction in various and unpredictable ways, exhibiting as well the following kinds of deep-rooted biological tendencies: mocking and petty criticism, caricatural misrepresentation (as in expressionist art), a secondary rather than a primary social spirit (organized as it is by rational criteria), dogmatic attachment to religion, which is reduced to ritual expression of formulas for living. Such a type cannot make good friends (they tend to dissolve relationships), they suffer from diminished vitality (being feminine and weak). This type is unfit to live among Germans who have already been corrupted by their spirit. People of this type must be removed from German public and economic life, even if it means "borrowing" their own cruel methods to do that. See Else Frenkel-Brunswik, "Environmental Controls and the Impoverishment of Thought," in Carl J. Friedrich, ed., *Totalitarianism* (New York, 1954), pp. 189–194.

57. Eliot's comment of 1934 is quoted in a relevant context by Harry Trossman, "After 'The Wasteland': Psychological Factors in the Religious Conversion of T. S. Eliot," *International Review of Psychoanalysis*, vol. 4, no. 3 (1977):303.

58. Hitler commented on modern art in *Mein Kampf*, for example, p. 235; the Vienna reference is in *Hitler's Secret Conversations*, p. 40; the reference to disintegration is in Waite, *The Psychopathic God*, p. 38. The art is evidence of an evil, uncontrolled, undisciplined, psychologically starved people, demonic and favoring darkness. The Jew is empty and is constantly seeking to fill himself up (bloodsuckers, parasites, maggots in a rotting body, etc.). "Has it not occurred to you," Hitler asked Rauschning, "how the Jew is the exact opposite of the German in every respect, and yet is as closely akin to him as a blood brother?" (See Steiner, *Power Politics and Social Change*, pp 111, 149). Hitler's images of the Jew, of course, are autobiographical statements, hence the "tantalizing aspiration to zero." The key image, however, is formlessness, chaos, disorganization, dissolution, fragmentation of the self. This involves problems of ego and reality, different from

the kind of problem suggested by sexually derived intrapsychic problems. The former range of problems leads to sociology, the latter does not. In any event, the Nazis constantly returned to the problems of form and order, to the state-forming will of the Aryans and the disintegrative will of the Jews: the Reich of the Teutonic soul rising against the anti-Reich of Jewish–Bolshevik chaos. The fear of chaos and disorder was the central dilemma felt by the Nazis. The image of the seducer-Jew lusting after blond virgins was not nearly as prominent or effective as the chaotic Jew. Needless to say, the state-forming will of the Aryan did not achieve expression in Germany until 1871, and then in the form of a tension-filled compromise. In less than 40 years that state was replaced by the Weimar Republic, which was not favored by Germans and was replaced in turn by the Nazi regime. The Nazi regime, on the Germans' own standards, was a disaster that culminated in the renewed division of Germany.

59. Quoted in Mosse, *The Crisis of German Ideology*, p. 296.
60. Quoted in Heyer, *Der Staat als Werkzeug des Bösen*, p. 120.
61. Mosse, *The Crisis of German Ideology*, p. 315. Hitler said, for example, that in Germany every judge must resemble every other judge, even in his physical appearance. *Hitler's Secret Conversations*, p. 87.
62. Himmler said in his speech to SS leaders at Posen, October 6, 1943, that the massacre of the Jews was a page of glory in their history, though it was a page that could never be written. Odilo Globocnik, by contrast, thought that later generations would not be so weak and feeble to fail to appreciate what the Nazis had done, that the Nazis had had the courage to complete this gigantic task of extermination. See Werner Maser, *Hitler: Legend, Myth and Reality* (New York, 1971), p. 247.
63. See, for example, Rudolf Binion, *Hitler among the Germans* (New York, 1976), pp. 33–34, 94–95, or U. D. Adam, *Judenpolitik im Dritten Reich* (Düsseldorf, 1972), pp. 312–313. On Goebbels, see Louis P. Lochner, ed., *The Goebbels Diaries* (London 1948), p. 103.
64. Ackerman, *Heinrich Himmler als Ideologe*, pp. 155–60; Hofer, *Der Nationalsozialismus: Dokumente*, pp. 85, 114.
65. George Steiner, "The Hollow Miracle," in *Language and Silence: Essays, 1958–1966* (London, 1967), pp. 122, 130. "The term 'scheisse' constituted the root for a series of hybrid expressions, such as 'scheisskommando'—the name given to the labor corps in Auschwitz and in other camps." Nachman Blumenthal, "On the Nazi Vocabulary," *Yad Vashem Studies*, vol. 1 (1957):58. Hitler justified himself by reference to Turkish treatment of Armenians, American treatment of Indians, and Stalin's treatment of his own population—the figure Hitler had was 13 million killed by Stalin. But whatever can be said about precedents, the alimentary fantasy was a unique German contribution.
66. Space is one of those venerable concepts that had a past in Germany and elsewhere, with an immediate geopolitical reference in the 1920s and

1930s. The concept was also susceptible of a number of interpretations. But the Nazis were singularly and peculiarly preoccupied with space. They devoted academic institutes to the study of space and they interpreted the world in terms of problems suggested by space. Thus, it was held that Russia had no history, only a geography, that is, Russia did not exist in time, only in space. In her dictionary of National Socialist usages, Cornelia Berning cites 14 compounds involving the concept of "space," and she did not get them all. There are no entries in her dictionary under the concept "time." Cornelia Berning, *Vom 'Abstammungsnachweis' zum 'Zuchtwart'—Vokabular des Nationalsozialismus* (Berlin, 1964), pp. 160–61; Max Weinreich, *Hitler's Professors* (New York, 1946), pp. 72, 127, 209; Klemperer, *LTI*, p. 218.

67. Rauschning, *Hitler Speaks*, p. 274.

68. The Nazis tried to develop their own high-level ideology centered on the primacy of affect, struggle, communal integration, and so forth. But this never became internalized morality for most of the population and the language was employed out of convenience, fear, opportunism, or it was construed in more traditional terms provided by the culture, as with the ideas of service, discipline, and sacrifice. Hitler more or less assumed this would be the case, as I have noted. On Nazi language, see Klemperer, *LTI*, pp. 49, 63, 97–100, 103, 161, 216–218; Dahle, *Der Einsatz einer Wissenschaft;* Berning, *Vom Abstammungsnachweis' zum 'Zuchtwart';* Blumenthal, "On the Nazi vocabulary," pp. 49–66; and Shaul Esh, "Words and their Meanings," *Yad Vashem Studies,* 5 (1963):133–167.

Index

A

Abel, Theodore, 28
"Absolute realism," 90
Acting out, 105
Affective processes, 8, 31–32, 51–57, 63–64, 68, 76 n.47, 78 n.52, 104, 106–107, 110 n.22, 116 n.39, 120, 128–131, 155 n.31
Alt Ausee, 151
Arendt, Hannah, 7
Artisans, 122
Aryans, 15–16, 41 n.45, 85–86, 89–90, 94–95, 102, 125, 133–135, 137, 141, 144–145
Asiatic chaos, 144
Austro-German culture, 84, 87
Autarchy, 133–135, 149
Authoritarian Personality, xv, 27, 29, 45–46 n.70, 106

B

Babylon, 114 n.27
"Background of Safety," 107
Baeck, Leo, 16
Baeumler, Alfred, 34 n.18
Bastardization of races, 144
Bavarian People's Party, 120
Beck, Ludwig, 14, 41 n.41
Benn, Gottfried, 2–6, 10–11, 13, 28, 32, 33 n.10, 34 n.16, 158 n.56
Berlin, 114 n.27, 143
Berning, Cornelia, 161 n.66
Binding, Rudolf, 13, 32, 40 n.40
Bismark, Otto von, 65, 68, 153 n.56
Blitzkrieg, 9 n
Blood poisoning, 144, *see also* Nazism
Blüher, Hans, 45 n.69
Bolshevism, 3, 25, 92, 95, 121 n, 137, 141, 144, 154 n.30, 156 n.38
Book-burning episode, 6, 13, 130
Bormann, Martin, 111 n.22, 125, 155 n.38
Bourgeoisie, 93–94, 103, 109 n.14, 109 n.17, 122, 124, 126, 132–133, 152 n.12
Boxheim documents, 22
Bracher, K. D., 34 n.17

163

INDEX

Braun, Otto, 64, 70
Buber, Martin, 22–23
Buck, Pearl, 63
Bullock, Alan, 96

C

Carossa, Hans, 14, 76 n.36
Catholic bishops, 69
Catholic Center Party, 120
Catholic Church, 138, 141
Charismatic leadership, 80–82, 107, 108 n.2, 115 n.35, 149, 151 n.5
Christianity, 12, 15, 21, 66, 83, 92, 94–95, 125, 137, 141
Class structure, 47–50, 59, 119, 124, 132
Cognitive processes, 17, 23, 31–32, 54–55, 71, 81, 85, 86, 88–90, 110 n.22, 116 n.39, 120, 121 n, 131, 150
  concrete images, 86, 88–90, 103, 136–141, 145, 147, 153 n.13, 156 n.42
Concentration camps, 14, 128–129, 147–148
Conservatives, 15–21, 31, 124, 134, 142
Continuity, sense of, 21–29, 51, 70–72, 105, 118–119, 121
Corporation of German Universities, 14
Cultic worship, 135
Curtius, Ernst Robert, 61

D

Darré, Walther, 136
Defense mechanisms, 17, 18
Democracy, 3, 10, 38 n.31, 140
Deuerlein, Ernst, 81
Deutsche Rundschau, 14
Dicks, Henry V., 28–29
Dietrich, Sepp, 115 n.32
Discontinuity, 32, 137, 157 n.50
Domestication, 42 n.47, 92
Dora case, 105
Dostoevsky, F., 158 n.54

E

Eastern chaos, 144
Eckhart, Dietrich, 144
Edschmid, Kasimir, 129
"Egocentric thinking," 90
Eliade, Mircea, 125, 153 n.15
Eliot, T. S., 142
Entitlement, 20, 42 n.49, 100, 133, 135, 150
Equality, 140
Erikson, Erik H., 71–72
Europe, 97, 133, 143
Experimental literature, theatre, 143
Expressionism, 2, 4, 129, 143, 159 n.56

F

Family, 21, 66, 106–107, 133
Fantasy thinking, 139, 145–146, 155 n.36
Farmer–warrior, 127
Fascism, 36 n.31, 132
Final solution, 136
Form–formlessness, 4, 141, 143–144, 146, 159 n.58
Forssmann, Werner, 27
France, 95
Frank, Hans, 115 n.32, 135, 138
Frank, James, 26
Freemasonry, 141
French Revolution, 25, 40 n.37
Frenkel-Brunswik, Else, 106
Freud, Sigmund, 7–8, 15–16, 35 n.22, 51–53, 70–71, 89 n.92, 104–106
  affective processes, 8, 51–57, 104, 110 n.22, 118, 153 n.13
  cognitive processes, 89 n
Fricke, Gerhard, 25, 44 n.60
Fried, Ferdinand, 45 n.69
Fromm, Erich, 113–114 n.26
Führer-Gefolgschaft, 128

G

Genetics, 88, 141
Genocidal wish, 94 n, 146

German Nationalist Party (DNVP), 60–61, 65
Gisevius, Hans Bernd, 66
Globocnik, Odilo, 160 n.62
Goebbels, Joseph, 62, 111 n.22, 135, 146, 157 n.47
Great Britain, 95
Great Depression, 60–61
Greater German Reich, 126, 138
Greek Antiquity, 135
Grimm, Hans, 155 n.33
Gypsies, 140, 148, 151

H
Halder, Franz, 96
Hamburg-Altona, 61
Hanfstaengl, Ernst, 46 n.74
Hapsburg Empire, 143
Hauptmann, Gerhart, 14
Heidegger, Martin, 6–11, 28, 32, 33 n.10, 34 n.20, 35 n.24, 35 n.27, 66
Heisenberg, Werner, 62–63
Henry the Guelph, 135
Henry the Lion, 124
Heredity, 88, 137, 141
Heydrich, Reinhard, 158 n.56
Hielscher, Friedrich, 19, 42 n.48
Himmler, Heinrich, 109 n.20, 124–125, 127, 135–136, 158 n.56
Hindenburg, Paul von, 28, 65
History, 3, 126, 138–139
Hitler, Adolf, 1, 4–7, 9 n, 12, 20–21, 24, 26, 32, 38 n.31, 42 n.49, 57–61, 67–68, 80–103, 106, 112 n.23, 117–118, 121–122, 125–128, 131–132, 135–136, 138–139, 143–147, 150 n.56
  archaic strivings, 101
  anonymous soldier, xiii, 82, 84, 155 n.34
  charismatic leadership, 80–82
  cognitive processes, 85, 87–89
  "either-or," 98
  ideology, 82–83, 103, 117–118, 135

Jews, the, 83–91, 94–95, 136–137, 147, 159 n.58
Mein Kampf, 84–85, 89, 113 n.28
narcissistic dynamics, 97–103, 114 n.26, 115 n.35
nihilism, 93
omnipotent ideas, 95, 101
orator, 101, 131
paranoid dynamics, 95, 102, 106, 110–112 n.22
SA (Sturmabteilung), 61, 79–80, 127, 152 n.10
SS (Schutzstaffel), 127–128, 135–136, 138, 146, 148
world view, 92
Hitler cult, 131
Hoegner, Wilhelm, 26 n.
Holocaust, 8 n, 92, 107, 148
Hugenberg, Alfred, 61, 65

I
Ideology, 30–31, 80, 82–83, 93, 103–104, 107–108, 108 n.4, 117–123, 149, 151 n.3, 152 n.7, 8, 161 n.68
  defined, xvii n
Id mythology, 109 n.14
India, 134
"Internal emigration," 11
Internalized morality, 55, 88 n., 118, 151 n.3
Imperialist–chauvinist sentiment, 133

J
Jaensch, E. R., 159 n.56
Jesuits, 141
Jews, 12–16, 22, 24–25, 34 n.18, 40 n. 37, 41 n.45, 42 n.48, 43–44 n.59, 64, 77 n.48, 84–92, 95, 99, 111 n. 22, 135–136, 146–147
  Eternal Jew, 125
  "ferment of decomposition," 140, 157 n.51
  Hasidic Jews, 137 n., 140
  Hitler, Adolf, 84–91, 95, 99, 109 n.17, 136

Jung, C. G., 15–16, 41 n.45
Nazism, 12–13, 136, 139–144, 146–148, 152 n.13, 156 n.38, 157 n.51, 52, 158 n.56, 159–160 n.58
  Orthodox Jews, 137 n.140
  social change, 143
  symbolic function, 143
  World Jewry, 147
  Zionists, 86
Jung, C. G., 15–17, 19, 23, 41 n.44, 45
Jung, Edgar, 153 n.13
Jünger, Ernst, 10–11 n, 19, 36–39 n. 31, 45 n.69, 153 n.13

K
Kafka, Franz, 96
Kaiser Wilhelm Institute, 22
Kästner, Erich, 62
Kerr, Alfred, 158 n.56
Kessler, Harry, 61
Kittel, Gerhard, 13 n.
Klemperer, Victor, 26
Klepper, Jochen, 64
König, Joel, 26
Krebs, Albert, 46 n.74, 96
Krieck, Ernst, 34 n.18

L
Language, 17–19, 20, 23, 42 n.50, 87–90, 152 n.8
  cognitive processes, 17, 23, 88–90
  ideology, 20 n.83, 108 n.4, 119, 152 n.7, 8, 153 n.13 161 n.68
Lapland, 158 n.52
Leadership principle, 130, 132
Lenard, Philip, 34 n.18
Lessing, Doris, 53
Ley, Robert, 138, 144
Liberalism, 3, 123–124, 134
Linz, 86
Lipset, Seymour M., 47–48
Loewenberg, Peter, 48
Lorenz, Konrad, 19 n., 42 n.47
Luther, Martin, 68
Lüthgen, Eugen, 66, 130

M
Mann, Klaus, 33 n.6
Mannheim, Karl, 122–125
Marcus, Steven, 105
Marinetti, F. T., 145
Marx, Karl, 118, 155 n.35
Marxism, 3, 8, 21, 30–31, 34 n.18, 134, 151 n.4
Maschmann, Melita, 154 n.23
Metaphorical processes, 88–90, 137, 145, 153 n.13
Metternich, Prince, 14
Militarism, 117
Modern art, 1–2, 140, 143, 159 n.58
Moeller van den Bruck, Arthur, 45 n.69
Mussolini, Benito, 95

N
Napoleon, 68
Nationalism, 25, 123
National Biology, 12
National Revolution, 69
Nation-state, 122–123, 125
Naumann, Hans, 66
Nazism, 2–5, 8–10, 30–32, 57–62, 69, 79–80, 103, 117–122, 125–126, 135, 146–147, 150–151, *see also* Modern art; Time orientations
  affective processes, 128–131
  blood, 88–90, 92, 120, 124–126, 130, 135–137, 141, 143, 145, 148–149, 156 n.40
  conservative intelligentsia, 5–6, 15, 19–20
  hierarchy, 132–136
  ideology, 117–122, 161 n.68
  instinct, 130
  Marxism, 3, 8, 21, 30, 34 n.18, 121 n., 132
  space orientations, 126, 129, 148–149, 160 n.16
  violence, 126–128, 129, 133
  Volkspartei, 60
"New man," 9
Niekisch, Ernst, 45 n.69

Niemöller, Wilhelm, 14
November criminals, 58
Nuremburg, 138

O

Object loss, 104–105, 118, 123, 137, 149
Ontological dilemma, 8–9
Order–chaos, 4, 141, 144, 158 n.52
Ordensjunkers, 144
Ostow, Mortimer, 41 n.45

P

Paranoid dynamics, 110–112 n.22
Paris, 99, 114 n.27, 151
Parliament, 123
Parliamentary democracy, 3
Peasants, 122, 128
Pechel, Rudolf, 14
Piaget, Jean, 89 n., 90
Pinson, Koppel, 21–22
Planck, Max, 23
Platonic Ideal, 68
Poland, 94, 134
Polanyi, Michael, 22–23
Polish Corridor, 156 n.43
Popper, Karl, 51
Pornography, 86, 143
Potempa murder, 22, 153 n.13
Protestant Reformation, 68
Psychiatric epidemiology, 116 n.41
Psychoanalysis, 143
Pyramids, 138
Pustau, Erna von, 63

R

Racial death, 144
Rauschning, Hermann, 9 n., 92, 95–96, 113 n.23, 125
"Regression in the service of the ego," 17
Reichstag fire, 23
Ritual destruction, 148
Ritual optimism, 157 n.49
Röhm, Ernst, 80, 153 n.13
Rolland, Romain, 13

Roman antiquity, 135
Rome, 95, 114 n.27
Rosenberg, Alfred, 13 n., 135, 153 n.13
Routinization of charisma, 108 n.4
Russia, 134

S

Saint Paul, 141
Sauerbruch, Ferdinand, 14, 41 n.41, 112 n.23
Schemm, Hans, 154 n.30
Schmitt, Carl, 11
Schwarzschild, Leopold, 62
Second Reich, 93
Seidel, Ina, 14, 41 n.42
Shtetl life, 143
"Situational guesses," 23
Slavs, 99, 148, 151, 155 n.38
Social change, 139, 143, 148–149
Social Darwinism, 127
Social Democratic Party (SPD), 26 n., 48–49, 60
Sombart, Werner, 45 n.69
Soviet Union, 94–95
Spann, Othmar, 45 n.69
Speer, Albert, 113 n.23, 114 n.27, 150
Spengler, Oswald, 45 n.69
Spranger, Eduard, 13–14, 41 n.41
"Stab-in-the-back legend," 119
Stahlhelm, 65
Stalin, Josef V., 25, 97, 113 n.26, 114 n.28
Stapel, Wilhelm, 11–13, 39–40 n.35, 36, 37, 152 n.13
Stark, Johannes, 34 n.18
Star of David, 141
Strasser brothers (Gregor, Otto), 45 n.69
Streicher, Julius, 13 n., 86
Szilard, Leo, 22–23

T

Terrence, 145
Thalheim, Karl, 77 n.48
Third Reich, 2

Time orientations, 82, 120, 122–126, 137–139, 149, 153 n.15
Totalitarianism, 6, 30, 35 n.24, 46 n. 71, 107

U
University of Berlin, 14

V
Vatican, 94
Versailles Treaty, 21, 24, 34 n.18, 60, 81
Victim, 86, 90
Vienna, 84, 86, 143

W
Warrior brotherhood, 38 n.31, 128, 132
Weber, Max, 81, 81 n.

Wednesday Society (*Mittwochsgesellschaft*), 41 n.41
Weimar Republic, 25, 53, 56, 58–61, 71, 81, 100, 102, 160 n.58
Widukind, 135
Women, 155 n.31
Workers, 48, 69, 119
World War I, 25, 36 n.31
Wotan, 135

Y
Young Plan, 60
Youth, 63

Z
Zehrer, Hans, 45 n.69
Zernik, Charlotte E., 26
Zuckmayer, Carl, 31, 43 n.55

Frederic L. Pryor. The Origins of the Economy: A Comparative Study of Distribution in Primitive and Peasant Economies

Charles P. Cell. Revolution at Work: Mobilization Campaigns in China

Dirk Hoerder. Crowd Action in Revolutionary Massachusetts, 1765-1780

David Levine. Family Formations in an Age of Nascent Capitalism

Ronald Demos Lee (Ed.). Population Patterns in the Past

Michael Schwartz. Radical Protest and Social Structure: The Southern Farmers' Alliance and Cotton Tenancy, 1880-1890

Jane Schneider and Peter Schneider. Culture and Political Economy in Western Sicily

Daniel Chirot. Social Change in a Peripheral Society: The Creation of a Balkan Colony

Stanley H. Brandes. Migration, Kinship, and Community: Tradition and Transition in a Spanish Village

James Lang. Conquest and Commerce: Spain and England in the Americas

Kristian Hvidt. Flight to America: The Social Background of 300,000 Danish Emigrants

D. E. H. Russell. Rebellion, Revolution, and Armed Force: A Comparative Study of Fifteen Countries with Special Emphasis on Cuba and South Africa

John R. Gillis. Youth and History: Tradition and Change in European Age Relations 1770-Present

Immanuel Wallerstein. The Modern World-System I: Capitalist Agriculture and the Origins of the European World-Economy in the Sixteenth Century; II: Mercantilism and the Consolidation of the European World-Economy, 1600-1750

John W. Cole and Eric R. Wolf. The Hidden Frontier: Ecology and Ethnicity in an Alpine Valley

Joel Samaha. Law and Order in Historical Perspective: The Case of Elizabethan Essex

William A. Christian, Jr. Person and God in a Spanish Valley